The Thee Generation

The Thee Generation

Reflections on the Coming Revolution

TOM REGAN

TEMPLE UNIVERSITY PRESS
PHILADELPHIA

Temple University Press, Philadelphia 19122

Published 1991
Printed in the United States of America

The paper used in this publication meets the minimum requirements of American National Standard for Information Sciences—Permanence of Paper for Printed Library Materials, ANSI Z39.48-1984 ∞

Library of Congress Cataloging-in-Publication Data
Regan, Tom.
The thee generation : reflections on the coming revolution / Tom Regan.
p. cm.
Includes bibliographical references and index.
ISBN 0-87722-758-6. —ISBN 0-87722-772-1 (pbk.: alk. paper)
1. Life and death, Power over. 2. Animal rights. 3. Ecology—Moral and ethical aspects. 4. Feminism—Philosophy. I. Title.
BJ1469.R43 1991
179′.1—dc20 90-40633
CIP

To Claire and Bill

For your kindness,
with my love

To clarify thought . . .—to do this, strange though it may appear, might be a way of saving human lives.

—Simone Weil

The oppression that black people suffer in South Africa—and people of color, women, and children face all over the world—is the same oppression animals endure every day to a greater degree.

—Alice Walker

In the relations of humans with the animals, with the flowers, with all the objects of creation, there is a whole great ethic scarcely seen as yet.

—Victor Hugo

Contents

Preface

The essays collected in this volume represent some of my recent work in moral philosophy. Though they differ in many ways, they are the same in others. Each essay offers answers to important moral questions, and each attempts to find the right reasons for these answers. The reasons are as important as the answers. To use an analogy: 20 is the right answer to the question, If you divide 100 by 2 and subtract 30, what do you get? Someone who arrived at this answer by thinking that 100 divided by 2 equals 75, and that 30 subtracted from 75 equals 20, would give the right answer but not for the right reasons. In math, one's reasons matter.

The same is true in morals. Of course, moral questions differ in many ways from questions in mathematics. The analogy is only that: an analogy. Still, the analogy makes an important point. Just as one might give the right answer to a question in simple arithmetic without understanding why it is the right answer, so someone might give the right answer to a difficult moral question without understanding why that answer is the right one. To do moral philosophy at its best, then, is to understand why the right answers are the right ones. This is what I try to do in each of these essays.

These essays also share one other feature. The moral questions examined in each inevitably force us to consider the boundaries of the moral community. The dominant moral traditions of the Western world, both religious and secular, place these boundaries at the limit of our species—the species *Homo sapiens.* Human beings are "in," every other life form is "out." If there is a central philosophical theme uniting these essays it is that this arrogant humanistic tradition cannot be rationally defended. More than the human is morally considerable.

In reaching this conclusion I claim no originality. Many dissident thinkers, both past and present, hold the same position, but none for the same reasons I offer. And that is important. If what we seek are not only right answers but the right reasons for them, then there is room enough, in the reasons one gives, for possible originality.

Perhaps there is room elsewhere. Considered as a group these essays address a diverse audience, an audience seldom assembled in the same place at the same time. Environmentalists. Feminists. Research scientists. Theologians. Psychologists. Philosophers. Policy-makers. Preprofessional and other students. Legal theorists. The captains of agribusiness. Physicians. Defenders of the rights of children. Economists. Political scientists. Hunters and trappers. Advocates of animal rights. The general public. As I say, a diverse audience. But even among the diversity there is a certain sameness, both in the challenges I make and that others are making, and also in the growing confidence that these challenges themselves are the harbingers of fundamental social change.

There is a revolution coming—or so I believe—this time in the realm of shared values rather than, as already has happened because of advances in science, in our understanding of facts. If others have yet to see this, I might claim this "original insight" as my own. But here, truly, who-thought-of-what-first is trivial. It is the truth of what is thought that is important.

And the origin of the coming revolution *is* thought, not the barrel of a gun. The values that define the revolution are those reached by informed reflection, not synthetic highs, and its partisans are united not by sporadic moments of group ecstasy but by shared ideals shaped and defended by rigorous thinking, ideals made real in the choice of one's career, in the details of one's day-to-day life, and in the experience of one's being-in-the-world, for example.

For this is a revolution of the human spirit. The central questions shaping the revolution are variations on ancient themes: Who are we? Why are we? But the answers given are those of a newly sensed humility: The human is but one life form among many, and what distinguishes us from the larger community of life is not our power to subdue but our responsibility to protect. In the coming revolution an expansive ethic of service replaces the suffocating ethic of greed. The daily question is What do I have to give? not What can I get to keep? A new generation—The Thee Generation—is aborning. The value, if any, of the essays collected here depends entirely on the contribution they make to the birthing. My personal hope is that even when they go wrong, they might still be part of the revolutionary process. As a philosopher, my hope is that my efforts might encourage others to play the important revolutionary role for which our discipline prepares us.

Most of these essays originally were presented as invited lectures. Some are pedagogical, others theoretical. In some the terrain I cover is familiar to me, in others I am the proverbial neophyte. The essays themselves, in a word, reflect the disparate audiences I have addressed and chart my efforts to come to terms with the challenging ideas of others. When necessary I have edited the lectures with a view to lessening the sense of public address and, in the notes at the end of the book, I offer a brief sketch of the circumstances surrounding their original presentation. I am pleased to express my gratitude to all those who, by asking me to speak, forced me to think. Without their gentle prodding I might not have moved at all.

I wish to thank Lesley Ann Beneke and Debby Stuart for their expert editorial assistance, Meg Kennedy for her help in preparing the book cover and jacket, and Jane Cullen for her wise counsel and encouragement; the faculty and administration of my university—North Carolina State University—for their commitment to honoring the principles of academic freedom; and, most of all, my wife, Nancy, for helping shape the moral aspirations we share. What voice I have is ours together, not mine alone.

The Thee Generation

1 *The Thee Generation*

I write of a new generation, The Thee Generation. It is a generation of service: of giving not taking, of commitment to principles not material possessions, of communal compassion not conspicuous consumption. If the defining question of the present generation is What can I get for me? the central question of this new generation is What can I do for thee?

The "thee" to be served includes the handicapped and the poor, the illiterate and the homeless, the starving and the abused, those newly born and those soon to die. Race makes no difference. Sex makes no difference. Religion and nationality, these too, make no difference. All who stand in need are served, with those least able to help themselves receiving the greater share. In The Thee Generation, absolute vulnerability finds absolute protection.

This ethic of service-to-the-other is rooted in a reconceptualization of what the human person is. A shared sense of community replaces the void of individual estrangement. Only by acting for the other does one come to know one's self, not in isolation from the ties that bind each to all but in affirmation of them. Apart from such relationships the self is seen to be an empty shell, the word "I" the most impersonal of pronouns. The malign logic of Descartes's *cogito* is dead. In its place a new declaration is alive: *Ego vivo in civitate, ergo sum* (I live in community, therefore I am).

Dissolved are the arrogant boundaries of humanism. Possessed of the breath of life, members of The Thee Generation recognize their membership in the life community. Each thing remains what it is, and not another thing. But *what* each is, is now seen to be connected to all that is. Whereas previous generations saw a world consisting of discrete, isolated atoms of matter (or of mind), The Thee Generation sees interrelated, molecular communities of life.

This change in perspective is more than conceptual. The realization of human embeddedness in nature is a half-truth if limited to some abstract formula, as if all biotic connections could be reduced to the chemical kinship found in the carbon cycle. Human embeddedness is subjectively concrete, affording an immediate, living sense of what it is to be in the world. The landscape remains unaltered, but the perspective is changed. Fundamentally. Radically.

Along with an increased awareness of living-in-community with all life, members of The Thee Generation assume increased responsibility for life. In contrast to previous generations, who saw the alienated human as master, members of The Thee Generation see themselves as protector of the natural world. The question What can I do for thee? is asked not only of vulnerable, disempowered humans but also of the other animals.

The depraved instrumentality by which previous generations have fixed the value of these animals is replaced by the recognition of their inherent worth. That worth is neither respected nor served by allowing commerce in their flesh as food, for example, or by utilizing them as "models" of human disease. An ethic of domination, one that elevates the human to the status of master species, both accepts and encourages these and many other forms of socially sanctioned barbarity. But not an ethic of service. Such an ethic shakes the very foundations of Western civilization as we know it. In place of the pathological tradition of animal enslavement, The Thee Generation offers a philosophy of liberation, one that calls for freeing these animals from the yoke of human tyranny.

This same philosophy extends to the sustainers of life: the land, the waters, the air. These, too, are protected—these, too, liberated—from the ravages of individual and collective greed. The human walks gently upon the earth, and what once was plundered is now restored.

Familiar problems remain. Members of The Thee Generation are not naked noble savages; they do not abandon human civilization in favor of the traditions of ancestral nomads. Within the network of the evolving human lifeway sources of energy must be found, food and building materials secured, means of transportation developed, the needs of the elderly attended, the young educated. Today's moral, social, and political questions endure.

But not today's answers. And not today's accepted basis for finding them. In place of the false standards of human material prosperity and personal salvation The Thee Generation calls for community integrity and individual service. In The Thee Generation the rights of one person end not only where those of another human's begin but also where the interests of the life community are threatened. The limits of individual liberty are reassessed. If educational practices encourage and reward acceptance of human isolation and domination, they are changed. When the few seek economic power at the cost of massive destruction to the life community, they are stopped. Since pornographic depictions of women subordinate women as a group, they are prohibited.

Because The Thee Generation is still emerging, it today has little by way of a settled philosophy. Indeed, perhaps more than anything else, the *philosophical foundations* of this emerging worldview—a worldview that tolerates even as it is not hindered by the latest "postmodern" intellectual fad—must receive the most intense exploration. And it is this exploration, not the franchised menu of ideas currently dominating academic philosophy, that define philosophy's real promise in the decades ahead and its debt to the centuries behind. In the best sense, then, philosophy's mission remains the same, as do those of the several sciences, art, religion, and other human creative endeavors. Again, only the perspective has changed.

That perspective already is changing, and the work of philosophers already is helping change it. Each of the insistent voices has its own vocabulary, each its particular agenda. Feminism.

Deep ecology. Animals rights. The differences are many, but the aspirations are the same: To rediscover who we are, and what we can (and ought to) be. A revolution is under way—a revolution of the human spirit. All around us the weight of dead theories and decaying institutions is being cast off.

I have written of The Thee Generation as new. Yet in some ways it is ancient. Like St. Augustine's City of God, which has no particular location, The Thee Generation has no specific temporal place. Some members have lived and died. Others live now. Many more are yet to be born. All are united across time by the bonds of shared ideals, communal aspirations, and an unshakable conviction: The day *will* come when their numbers and influence will be sufficient to cause cultural change, to save not only the whales and the planet but ourselves. Faith demands this. And optimism? Optimism finds cause to celebrate the early signs of restlessness among today's youth, only now beginning to rouse themselves from their parents' dogmatic slumbers: the *next* generation!

One thing is certain. Whatever hope there is for the flourishing of life, within and beyond the human family, requires the demise of those ideas that have brought us here. The ethos of avenging angels is past. Ours is an age when benevolent assassins are asked to bury false ideologies. If philosophers have a future, it is this.

2 *Living as if Life Matters*

Plato was one of the greatest of all the world's philosophers. The twentieth-century English philosopher Alfred North Whitehead even went so far as to declare that "after Plato, the rest of philosophy is footnotes"—and this said of a thinker who died in 346 B.C.E.

Whatever Plato's true place in the pantheon of philosophical greats, there are two features of his thought of special relevance here. The first concerns his uncommon ability as a storyteller, the second his distinction between right opinion and understanding. I shall comment on both in the order just given.

Plato ranks among the most rigorous of thinkers, one who, in the person of Socrates throughout all of Plato's writings, relentlessly pursues the last drop of logical truth. Yet Plato has a keen sense of the limits of rational argument and, relatedly, an astute appreciation of human psychology. If the heart has its reasons that reason does not understand, then Plato must be credited with a profound appreciation of this truth. We glimpse this whenever he allows the argument to tarry and permits Socrates or another person in one of his dialogues to wax imaginative: to tell a story. It is one such story I particularly want to draw to your attention.

The story is called the Allegory of the Cave and it is presented by Socrates approximately three-quarters of the way through Plato's most famous dialogue, *The Republic.* We are to imagine the following. A group of people have lived all their life in a cave. To make matters worse, they have been shackled in such a way that only one wall of the cave has been visible to them. And this is important in the story. For behind them, out of sight, the cave is alive with activity. Unseen people work as keepers-of-the-fire, insuring that the cave is first illuminated then darkened in a rhythmic fashion, much as for us day follows night, and night day. Still others busy themselves creating lightweight images of things made to resemble such objects as a house or tree, a child or dog, a flower, a table, a chair. Still others, finally, perform the function of image-carriers. While the cave is illuminated they carry various images of things over their heads, their arms stretched to their limits. All this is going on in the cave, as I have said, yet not one of the cave's prisoners has the slightest clue. Positioned as they are in the cave, their daily, shared experience consists of this: On the wall before them they see the shadows of the images of things cast by the fire.

Understand, they have never seen a real tree or child, never even seen a thing made to resemble a tree or a child; all they have ever seen are the shadows of such things cast on one wall of their cave. And yet for these people these shadows are reality. The real world. The truth. Were any one of them ever to want to be reassured that he or she is right, all that person would need do is ask the other prisoners. Each knows what is real and true, and each agrees with the rest.

That is part of the Allegory of the Cave. I shall return to the story's conclusion momentarily. But first I want to say something about the second feature of Plato's thought mentioned earlier—his distinction between right opinion and understanding. I can illustrate this distinction with examples.

George says, "The sun will rise in the east tomorrow," and what George says is true. George, that is, has right opinion. But understanding? Not if George is ignorant of the laws of celestial mechanics, gravity, and the like. Without knowing *why* the sun will rise in the east tomorrow, George's right opinion falls short of understanding. And a similar diagnosis applies to every similar case—for example, Barbara's right opinion to the effect that a piece of steel will sink in water while a piece of wood will float. Just as in George's case, Barbara's correctly believing what will happen does not mean that she understands why it will. To put the general point at issue here in the most general terms, understanding involves knowing why what is true, is true; to understand something, it is not enough merely to believe what is true.

Viewed in the light of the distinction between right opinion and understanding, the prisoners in the cave are in very bad shape. Not only do they lack understanding; they lack right opinion. They do not have the foggiest idea that the "truths" upon which they agree concern the shadows of images of things. And for as long as they all are shackled in their chains not one of them will ever know this.

Ah, but Plato does not have his well-deserved reputation as a storyteller for no good reason. He invites us to imagine that one of the prisoners—a man, as it happens, but today we are (or at least we should be) willing to imagine it is a woman—a man becomes freed of the shackles that heretofore had bound him and, once liberated, sets out on a solitary journey of discovery. First to be found are the images of things: one the shape of a table, another the shape of a chair, and so on. And then there is the fire, whose brightness is both frightening and inspiring. We are to imagine moments of real excitement: "So this is what a chair is! This, a table! This, the source of light!"

But the quest does not stop there. The liberated one notices a passage leading out of the cave and, summoning his courage, embarks on his perilous journey. Outside, the sun's light is overpowering and it takes several minutes for his eyes to adjust. At first he is only able to observe the reflected images of things—the image of a tree on the stilled surface of a lake, for example; then the shadows of these things; then, finally, the things themselves: an actual tree, a flower, houses, people. What a revelation! "So *this* is what reality is like!" we can imagine him saying to himself. "The truth is not to be found in the shadows on the wall of the cave, not in the images people make, not even in the images or shadows of these things, but in these things themselves. Before I lived in falsehood; now I have glimpsed the truth."

Glimpsed it, yes; but only that. For there is much that the liberated one does not know, much to be learned and understood. And therein lies his dilemma. For were he to accede to his desire to know more he would be more likely to succeed if he remained outside the cave—outside the world of illusion. Were he to make this choice, however, his former fellow prisoners almost certainly would be destined to live the rest of their lives in a universe of false knowledge.

What to do? Plato imagines that the liberated one returns to the cave knowing full well that he faces incredulity and even ridicule when he attempts to convince the prisoners that what they think is true actually is false, that what they think are real things actually are shadows.

These are the main outlines of the Allegory of the Cave. As an allegory the story means more than it says. An allegory, after all, is a representation of one idea in the guise of another. So what is the story *really* about? This is the question I ask my students when I relate the Allegory of the Cave, as I always do, during my first lecture in all my philosophy classes. To answer this question it is helpful if we first ask another one, namely, Who are the prisoners in the cave?

My students offer a variety of answers, ranging from Madison Avenue ad agents to liberal democrats. But sooner or later one of them gets it right: *Ordinary human beings* are the prisoners in the cave. The great mass of humanity lives in a shared universe of ignorance, blissfully content to accept as true what they have been taught by the culture in which they were born and raised, uncritically digesting the received opinions of their time and place and circumstances, daily finding congenial confirmation of their view of the world in the happy chorus of similar opinion voiced by those around them. It is the great mass of humanity that is represented by the prisoners in the cave. Or so it seems to me.

Who, then, is represented by the liberated one—the prisoner who escapes, realizes the depth of his ignorance, and, against all odds, returns to the cave, there to attempt to free others from the chains of their comfortable ignorance? I do not think there can be any doubt what Plato thinks the correct answer is. The liberated one is the philosopher (or, to speak more accurately, the true philosopher—for there are many imposters who, in their cleverness, can seem to be philosophers when in fact they are not). And if we ask for the inspiration of Plato's lofty representation of the true philosopher we find it in the person of Socrates. For it was Socrates who never tired of professing his ignorance and who never ceased to be amazed at the false assurance of knowledge he found among most of his contemporaries. When, in response to the question Who is the wisest man in all of Athens? the Oracle at Delphi replied that it was Socrates, Socrates interpreted this to mean that his wisdom consisted in his knowing that he did not know, whereas the ignorance of others consisted in their thinking that they did.

Now, it is significant that the Allegory of the Cave is presented in Plato's most important work in the area of moral and political philosophy. The central question in *The Republic* is What is the ideal state? or, alternatively, What is the best possible form of government? To answer this question requires that we plumb the depths of an even more fundamental one, namely, What is justice? This is required since, whatever the details of the ideal state happen to be, any state that lacks justice must be less than the ideal one.

For present purposes how Plato answers these two questions is less important than that he presents the Allegory of the Cave in the very same context in which he discusses them. His views therefore seem unmistakable: Most people are as far from understanding what makes some laws just, others unjust, and some acts rights, others wrong, as the prisoners in the cave are from understanding what a table or chair is. Granted, the great mass of humanity might *think* they know these things, just as the prisoners in the cave think they know what is true; and granted that, because of this uncritical self-assurance, most people, like most prisoners, will resist or abjure any and all efforts to wrest them from their comfortable illusions; nevertheless, it remains true that the presence of "right opinion" by itself (when this is present) no more gives understanding to ordinary men and women in the streets than it would to prisoners in the cave; that if understanding is to be acquired one must understand *why* what is true, is true; and that the only hope we have of finding such understanding is to follow Socrates's example: We must think *philosophically.* And this is true whether our interest lies in understanding the basic nature of the physical, or of the moral, universe.

Sacred and Secular

It is the moral universe I wish to explore. I shall do this by first asking two related questions: What makes right acts right? And what makes wrong acts wrong? To answer these questions it is not enough to give examples of each—not enough, that is, to say "Rape is wrong," "The murder of the innocent is wrong," "Racial and sexual discrimination are wrong," for example. For even if we concede, as I think we should, that people who believe these propositions believe what is true and thus have "right opinion," this is not the same as understanding. To understand why murder is wrong, to take just one example, we need to do more than merely to believe (rightly) that it is.

When we press these questions (What makes right acts right? What makes wrong acts wrong?) we find a variety of responses among ordinary people (different responses in different caves, as it were). Some of these responses are religious, others secular. Responses of the former (the religious) kind offer answers to our questions by making reference to the sayings or teachings of one or another sacred text or religious figure (God or Jesus or Muhammad, for example), whereas responses of the latter kind (secular responses) dispense with any and all such references. Since my interest here is not completeness I shall characterize and critically assess only a few examples of each.

The Divine Command Theory

When asked why murder is wrong, many people who connect their moral beliefs with their religious convictions reply, "It says so in the Bible," or "It's one of God's commandments—'Thou shalt not kill,'" or "God forbids it." Of course people who do not share these religious convictions are likely to dispute them; and there is no denying the relevance of this line of debate. It is important to realize, however, that the adequacy of a religious-based ethic can be challenged without challenging belief in God.

One way to do this is by first asking for additional clarification about the supposed connection between what God is said to command and the wrongness or rightness of various actions. In the particular case of murder, for example, there are two possibilities. The first is that murder is wrong *because God forbids it;* the second is that God forbids murder *because it is wrong.* Both alternatives can be shown to be unsatisfactory. And since one must choose one *or* the other, if one wishes to base one's ethic on God's commands, the result is that no such ethic is satisfactory.

To make this clearer consider the former alternative—the one that says murder is wrong *because God forbids it.* This option has quite a few forbidding consequences of its own. To begin with, if we are asked to suppose that God's act of forbidding murder is a historical event—if, that is, God performed this act at a particular time and place (say, when God gave the Ten Commandments to Moses)—then the view under examination implies that murder was *not* wrong prior to the time at which God performed this act. For many thousands of years, in other words, we are being asked to suppose that murder was not wrong, but once God issued the prohibition, it became wrong and has remained wrong ever since.

This is a radical distortion of the foundations of morality. What is legal or illegal can change in the way we are imagining, as when, for example, a benevolent dictator proclaims that it shall no longer be legal to buy and sell slaves. In a case like this we can say that what is now illegal formerly was legal, and that this change in legal status is the result of an action performed by the benevolent dictator at a particular time. But what is moral or immoral is (and must be) quite different. If murder is morally wrong today, then it was no less morally wrong in all the yesterdays you care to enumerate, and any ethical position that implies the opposite cannot command our rational assent. Since one way of interpreting the supposed connection between the wrongness of murder and what God forbids does imply the opposite, this interpretation will not pass muster.

In response it may be said that God has always forbidden murder—perhaps, even, that God has done so eternally. On this interpretation, then, unlike the first one, murder did not become wrong when, say, God gave Moses the Ten Commandments; on the contrary, murder has *always* been wrong because God has *always* forbidden it.

This second interpretation avoids the fatal implication of the first one but only at a no less damaging price. *On what basis,* we must ask, is it being claimed that God has always forbidden murder? The most plausible answer makes reference to God's omnibenevolence—God's supreme or perfect goodness. If we assume that God has such perfection, then we might infer that this includes *moral* perfection. And if we assume that a morally perfect being would always forbid what is wrong, then, since God does forbid murder, murder is now and always has been wrong.

Whatever initial plausibility this interpretation has evaporates under the gaze of careful reflection. If morality really does depend entirely on what God commands, then *God's* moral perfection must depend on this too. And this would make God's moral perfection result from the fact that God commands it, which is absurd. Granted, there may be a God. Granted, this deity may be morally perfect. What is not to be granted is that the moral perfection of any deity can be the result of a command. Even if we suppose that God was able to create light by saying "Let there be light!" not even God could be made morally perfect by saying "Let me be perfect!"

Faced with these formidable objections people who wish to base their moral beliefs on their religious convictions might have recourse to the second alternative mentioned earlier. Instead of maintaining that murder is wrong *because God forbids it,* they might instead say that God forbids murder *because murder is wrong.* But once this is said any semblance of a religious-based ethic disappears. For if God forbids murder *because it is wrong,* then what *makes* murder wrong must be independent of anything God commands, in which case the answer to our questions—What makes wrong acts wrong? And what makes right acts right?—must be explored independently of anything *any* religious figure has commanded or forbidden.

Now, more could be said about the particular religious-based ethic I have been critically assessing (the Divine Command Theory, as it is called), and it is also true that there are other ways in which such an ethic might be attempted. For present purposes, however, enough has been said to suggest the sorts of problems such attempts are likely to encounter, even assuming that there is a god, that the deity is morally perfect, and so on. The general conclusion hard thought reaches, I think, is simply this: People who base their beliefs about moral right and wrong on their religious convictions may have right opinion—as when, for example, they believe that murder is wrong. What they lack is understanding, in the sense that their religious convictions fail to explain why what is wrong, is wrong, and why what is right, is right. To this extent, and for this reason, such persons, assuming they rest content in their ignorance, are "prisoners in the cave."

Cultural Relativism

This same diagnosis applies to many people who keep their moral beliefs separate from their religious convictions (if, indeed, they have such convictions). These people understand morality in secular rather than religious terms. Among the many possible secular options one is *cultural relativism.* In place of the moral authority religious believers attribute to the word of God, cultural relativists put the majority voice of a culture at a given period of its history; and in contrast to the religious authoritarians, who find in God's commands the one true morality for all people, at all times, and in all places, cultural relativists believe that morality is relative in fundamental ways. For what *is* right, and what *is* wrong, they think, can—and often do—vary from one culture to another at the same time, and within the same culture at different times.

When we ask our questions of cultural relativists, therefore, their general answer is, "It depends." More specifically they will say something like the following. "What makes right acts right, and wrong acts wrong, are the beliefs and attitudes of the majority of the culture in which the question is asked, at the time one asks it. Thus, for example, it is wrong in the United States at this point in time to eat one's deceased parents, but this same thing has been deemed right in other cultures. Again, it may be laudatory within Iraq for Iraqi terrorists to kidnap or kill people loyal to American interests, while Americans, of course, condemn these very acts. From our (that is, the cultural relativists') point of view, *both* parties are correct, *neither* is mistaken."

Cultural relativists are not fools. Indeed, they tend to think that the real fools are the people who resist their position. These people are *ethnocentric:* They think the beliefs and values of the group to which they happen to belong are the *only* true beliefs and the *only* true values for everyone. These people therefore have a marked tendency to be morally intolerant of others. In the face of such intolerance, cultural relativism appears to be far less arrogant, far more humble. "Those who are ethnocentric," we can almost hear them say, "really *are* prisoners in the cave."

Well, perhaps they are. Still, we need to ask whether the amount of tolerance encouraged by cultural relativism is too much of a good thing. It seems that it is. Do we really want to be unable to make *cross-cultural* moral judgments? Think of what this prevents us from doing. We cannot say that the Nazis were wrong to exterminate six million Jews. And we cannot say this, given cultural relativism, because what is right and wrong for us is just that: right and wrong *for us.* Our standards are not everyone's standards. Provided only that the majority in Germany during the Third Reich approved of the "final solution," what the Nazis did was morally proper.

Few there are who will be prepared to accept a view that has this as one of its implications. And this is not the only unsavory outcome. Consider slavery. If the majority of Americans alive in, say, 1860, approved of keeping black-skinned human beings in bondage, then doing so was perfectly all right, according to cultural relativists. That most of us today think differently does not show that these earlier Americans were mistaken.

One would have thought otherwise. One would have thought that slavery was wrong back then, just as the murder of innocents by terrorists is wrong now, no matter how many people think otherwise. If it is possible for the majority to be mistaken about some matter of fact, as when most people thought the earth was flat, then why is it not equally possible for the majority to be mistaken about some matter of value? Thoreau's enduring image of the moral dissident, the solitary person who marches to "a different drummer"—surely this image cannot be dismissed simply on the grounds that the lone marcher speaks as one against many. In asking each of us, as individuals, to accept the error of our ways if our judgments of value conflict with the majority in our society, cultural relativists ask too much: They ask us to surrender part of what it is to be human, namely, our ability to think and judge for ourselves. Indeed, unless we assume something like this we would lack any reason or incentive for critically exploring the values of the culture into which we have been born and, if and when we find these values deficient, abandoning them—even in the face of social ostracism or ridicule. Cultural relativism, it seems, cannot give us the answers we seek. It allows us no way out of our cave.

Subjectivism

A third option, different from and arising out of the two discussed to this point, merits our consideration. This view is known as *subjectivism* because it emphasizes the independent judgment of each person—each subject (hence the name subjectivism). It is not God's commands that determine what is right and wrong. Nor is it the majority opinion current in one's culture. No, it is *what each person thinks,* or *what each person feels* that determines this. Let me explain this position more fully.

According to subjectivism some acts really are right, others really are wrong. What makes an act the one or the other does not depend on some feature of the act in question, but depends, rather, on how each individual person feels about them. In the case of those acts that are right according to one person (let us call her Nancy), Nancy feels some positive emotion: She likes, prefers, or approves of these acts, for example. And the same is true in the case of any other person—Ronald, say. When Ronald thinks that an act is right, he, too, feels favorably about it. As for acts judged wrong, subjectivists in ethics say the opposite. In this case each person has some negative emotion—dislike, disapproval, and so on.

What, then, according to subjectivists, does it mean to say "Murder is wrong"? Well, when Nancy says this, what it means is that she (Nancy) disapproves of murder; it is *the fact of her disapproval* that she states when she says this. However, the fact Ronald states is importantly different; when Ronald says this it is *the fact that he (Ronald)* disapproves of murder that is stated.

Subjectivism is as rationally flawed as any ethical position can be. To understand why, consider its implications. Suppose Nancy is sincere when she says "Abortion is wrong." If she is, then Nancy cannot be mistaken. And the same is true in every other case: Whenever Nancy says that something is wrong, and whenever she says that something is right, she cannot be mistaken, provided only that she is sincere.

But a person who cannot be mistaken in this way—one who always and only says what is true—is morally infallible. Nancy, therefore, is morally infallible. And this is absurd. For whatever else we may want to say about moral judgment we at least do not want to say that Nancy (let alone Ronald) is morally infallible.

Since the view we are currently considering—subjectivism—does imply moral infallibility, this view cannot be right. We are not being doctrinaire, therefore, but are instead only being fair to the truth when we say that people who are comfortable in their subjectivism are no less prisoners in the cave (though perhaps a different cave) than those who are comfortable in tying their moral beliefs either to their religious convictions about what God commands or to the prevailing values of their culture. Like the latter adherents, the former may have right opinion when it comes to the wrongness of murder or even the wrongness of abortion. But in concert with the latter, subjectivists in ethics lack understanding: They do not know why what is right or wrong is the one or is the other.

Exiting the Cave

Alternative secular answers to our central moral questions as these occur to the great mass of humanity are numerous and varied, so my brief critical examinations of cultural relativism and subjectivism only scratch the surface. Still, the stark deficiencies of these views suggest that we are unlikely to find anything approaching a satisfactory answer to our questions if we look for them among the ill-considered beliefs of the majority of people. Answers (if there are any) must be looked for elsewhere. And if we ask where, then (mindful of Plato's Allegory of the Cave) we can answer, "In the thought of those *philosophers* who have dared to venture outside the cave and, upon returning, have shared their vision with us. Surely here is where we will find the answers we seek."

Alas, the situation is not as cheerful as this prescription might suggest. The plain fact is, different philosophers have returned, not with the same, but with different visions. *Many* different visions. If we think of philosophical theories of ethics as cars, we can say that there is not just one, there are many brands, and—among the brands—many models, and—among the models—

many options. Even to get an informed sense of the *possibilities* is a formidable task. How much more daunting, therefore, is the challenge to choose the best of the lot—that theory which, all considered, is closest to the truth. To do this may take a lifetime—and then some. But let us press on as best we can.

The Moral Community

Suppose we imagine the moral community circumscribed by a circle. Individuals inside the circle are members of the moral community, individuals outside the circle are not. Those inside the circle, by virtue of their membership in the moral community, are entitled to a kind of consideration denied those outside. Of the former, but not the latter, we may say, "They are morally considerable."

To say that an individual is morally considerable (or, alternatively, is a member of the moral community) implies at least the following: Various principles recognized as moral principles in the community, principles whose purpose is to protect the members of the community, should function to protect this individual. Thus, for example, if the prohibitions against rape and murder are recognized as morally binding within the community, then each appropriate individual within the community should be protected against rape and murder. Normal sanctions (for example, various fines and other punishments) should be applied to anyone who violates these prohibitions, whatever the victim's identity. In this way, although of course in others as well, all potential victims within the community are afforded the protection of the community.

Moral communities can be investigated in different ways by different disciplines. One way to make these differences clearer is to distinguish between *descriptive* ethics and *normative* ethics. Here is the way I understand this distinction.

The goal of a person doing descriptive ethics, as the name suggests, is accurately to describe various features of the moral life, including in particular the conditions that qualify one to be a member of this or that identifiable moral community. Thus, descriptive ethics is the sort of inquiry carried out by those who do empirical research into one or another aspect of the moral life. As traditionally conceived, descriptive ethics is the business of the social sciences—psychology, sociology, and anthropology, in particular.

The goals of normative ethics, by contrast, are to establish who *should* belong to *any* moral community, and what ethical principles *any* moral community *should* acknowledge. As traditionally conceived, normative ethics is the province of normative moral philosophy, the kind of endeavor we find in the work of such legendary figures in the history of philosophy as Plato, St. Thomas Aquinas, Immanuel Kant, and John Stuart Mill.

Whatever else we might want to say of these two ways of approaching ethics—descriptive ethics, on the one hand, and normative ethics, on the other—at least this much seems clear: They really are conceptually distinct. The descriptive ethicist is concerned with *what is the case* (whether it "ought" to be so or not), whereas the normative ethicist is concerned with *what ought to be the case* (whether it "is" so or not). Our interests here, as they have been throughout, are mainly of the normative variety.

Two Assumptions

In order to pursue these interests further I need to make two very general assumptions. The first is itself transparently, openly, avowedly normative in character. I assume that the ideal moral community will include some principles with which we are familiar—principles having to do with the importance of fairness, the value of honesty, the general demand to treat others with respect, and the more particular obligations not to rape or murder, for example. I realize that these are large assumptions to make and that, stated nakedly, bared of any supporting argument, cultural relativists might smirk at what they regard as ethnocentrism-on-parade. In another time and setting I might be able to quiet this suspicion. Here and now I can only hope that it does not seem unreasonable.

My second major assumption is not transparently normative (which does not mean it is not normative). I assume that the following general principle (which I shall call "the formal principle of justice") is a rational decision-making principle: In order for it to be rational to include one individual (A) in, and exclude another individual (B) from, the moral community, one must be able to cite a morally relevant difference between the two—between A and B, that is. And what I mean by "a morally relevant difference" is a difference on the basis of which it would be rational to include the one, and exclude the other, from the protection afforded by such principles as those I mentioned in my first (my avowedly normative) assumption. Perhaps the meanings of and the relationships between these two assumptions will be clearer if I use some examples.

To begin with, here are a few farfetched ones. Suppose that A is included and B excluded because A is taller than six feet and B is not. I think we all would agree that, thought the two do really differ in their height, this is not a morally relevant difference. For there is nothing in the existence of this (height) difference that could remotely make it rational to include A within the moral community's protection against murder and rape, for example, while excluding B. At the very most what we have here is an arbitrary prejudice that exalts the moral status of tall people (what we might call "tallism"). And analogous remarks would apply to anyone who exalted the moral status of, say, right-handed people ("rightism") and fat people ("fatism"). In every such case, a feature is cited (tall, right-handed, fat) in terms of which people really do differ but in no such case is this difference a morally relevant one.

These examples, as I have mentioned, are farfetched. No one (to my knowledge at least) ever has advocated tallism or rightism or fatism in a serious way. And yet these examples are close conceptual relatives of some actual prejudices that, especially when they have been institutionalized and socially sanctioned, have led to profoundly tragic consequences. I have in mind not only sexism but also the various forms of racism of which our history is so full. A person with ovaries and fallopian tubes differs sexually from me, just as Jesse Jackson and Fidel Castro differ from me in the pigmentation of their skin. But to suppose that differences of this kind qualify me, and disqualify them, for membership in the moral community is both a real and a historically tragic violation of the principle of formal justice.

Why Murder Is Wrong: A Hypothesis

The formal principle of justice takes us part of the way toward answering a specific question like What makes murder wrong? This principle entails that arbitrary double standards are not allowed. If it is wrong to murder Smith then, in the absence of a morally relevant difference, it is no less wrong to murder Jones. And this is true independently of what one or another religious figure or sacred text declares, independently of what the prevailing beliefs or attitudes of any culture happen to be, and independently of the beliefs or preferences of this or that individual.

As powerful as the formal principle of justice is, however, by itself it does not answer the particular question about the wrongness of murder or the more general questions about right and wrong we have been considering. How might these questions be answered? Let us consider the specific question about the wrongness of murder first.

Surely any plausible answer must rest on considerations about the value of human life, since if human life were devoid of value it would be strange—to put it mildly—to believe that murder is wrong. So, what *is* there about human life that makes it so valuable? Once again we are not lacking for possible answers, and once again some of them are secular, others religious. Among these possibilities one seems to me to be the best. Here, roughly, is how I see it.

Human beings not only have a biology, we have a biography—a life story that unfolds from the "inside," so to speak. We bring the mystery of consciousness to our separate lives and, what is more, the conscious presence each of us is fares well or badly. Food we need. And water. Clothing. Shelter. These are the minimum for faring well. Pain, anxiety, stress, loneliness, restrictions on our freedom—these factors, whether chronic or acute, subtract from our welfare, while pleasure, relaxation, a sense of social connectedness and acceptance, a generous helping of freedom—these factors add to our sense of well-being. But whether we are fortunate and are faring well, or unlucky and faring poorly, one thing is true: The life we are living is ours, of value to us, and it is not to be taken from us in the absence of very special reasons.

Some people—pacifists such as Gandhi, for example—maintain that it should not be taken from us for *any* reason. And possibly this is true. It seems more likely, however, that there are some situations, such as those of self-defense or defense of the innocent, where the taking of a human life is not wrong. But whether always wrong or sometimes right, the willful, deliberate, intentional taking of a human life at least is not always right, and one way of trying to say why it is not is suggested by my earlier remarks about our status as biographical beings. To continue this biographical theme, we can say that other people are not at liberty to end our life story because they are angry or jealous, for example, or because they or others stand to benefit as a result of our death. To end our life for such reasons is to make our story a part of their story, without our consent and contrary to the right to life that we possess, as the subjects of the story. That, as I say—in a sketchy way—is how I think the question Why is murder wrong? should be answered, and while I do not think that every question about right and wrong can be answered by appealing to the right to life in particular or to human rights in general, I do think many important questions can be answered in these ways.

Old Business, New Business

Two points deserve brief comment before concluding. The first is in the nature of old business. The incompleteness of my answer to the questions What makes wrong acts wrong? And what makes right acts right? is (or should be) obvious. For example, my claim that we have a right to life needs to be defended, as would any other similar claim—for example, an appeal to our right to liberty. Clearly, a lot of hard thinking remains to be done before we have anything like a full-blown theory of right conduct. Even so, there is one piece of good news: My account of why murder is wrong does not contravene the principle of formal justice. *Any* biographical being—regardless of race, sex, religion, nationality, or, say, intelligence—is included. My biographical answer, in a word, does not rest on or permit moral arbitrariness. Whatever else we might think of my answer, this fact certainly seems to be a point in its favor.

Second—and here we come to the new business—the biographical answer I favor has some important implications it is possible to overlook. All the biographical beings considered to this point have been human beings. Now, because the one thing they all have in common—the one characteristic in terms of which none differs from any other—is species membership, perhaps it will be thought that it is *this* characteristic that picks out a morally relevant difference between those who are, and those who are not, members of the moral community. As I understand this suggestion the following is a way to test its validity. Imagine that A is a human being of average age and intelligence, while B is a nonhuman primate also of average age and intelligence. The two (A and B) differ in many ways, certainly; there are many things A can do that B cannot, and there are a fair number of things B can do that A would find difficult or impossible. However, the two also share quite a few similarities. For example, both hear and see, are sometimes hot and sometimes cold, are sometimes tired, excited, angry, curious, afraid, and lonely, are sometimes pleased and sometimes are in pain.

Is it anthropomorphic to view nonhuman primates in these ways? I think not. As Darwin observed more than a hundred years ago, the mental life of nonhuman primates and other (what he called) "higher animals" differs from ours in degree, not in kind. Unless we want to abandon some of the major implications of evolutionary theory, not to mention the unanimous verdict of common sense, therefore, I think we must judge Darwin right, the weary charge of anthropomorphism wrong. Like us, these animals are biographical beings.

The differences between A (the average human) and B (the average nonhuman primate) obviously could and, in my view, should make a difference concerning some of the protections and entitlements each should be afforded, if each is recognized as a member of the moral community. For example, while A should be protected against discriminatory recruiting practices in higher education, there is no point in insisting that B should receive similar protection, and this for the simple reason that nonhuman primates lack the abilities to learn or teach calculus, chemistry, and Shakespeare, for example.

But while such differences surely make *some* moral difference, what about the similarities? Do they matter morally, too? If (as I assume we agree) it would be wrong to kill a human being just because someone else stands to benefit, then how can the ethics of doing this to the nonhuman primate rationally be judged any differently? For my own part I cannot see how the judgment in the latter case rationally can be any different than the judgment in the former, which implies that our imaginary B, and other nonhuman animals like B in the relevant respects, rationally must be recognized to be members of the ideal moral community (whether or not any actual community is today characterized by this recognition). And this finding in turn implies that it would be no less morally prejudicial to assume or insist that belonging to the species *Homo sapiens* is necessary for membership in the moral community, than it would be to assume or insist that being a certain height or weight is necessary. Put another way, "speciesism" is as much a moral prejudice as tallism or fatism. The real and tragic difference is, few (if any) people have acted on the latter prejudices whereas many people have acted—and continue to act—on the former one.

Here, then, we assuredly reach a conclusion that, if true, shows how fundamentally relevant moral philosophy is. Its relevance manifests itself not only in those much discussed questions of the ethics of insider trading, duties to one's employer, whistle-blowing, fraudulent advertising, client confidentiality, informed consent, organ transplants, a woman's right to an abortion, a fetus's right to life, and a patient's right to die with dignity. Moral philosophy can go to the very foundation of accepted social practices and challenge their deepest moral basis. The sorts of institutional and professional changes it can call for sometimes are more than reformist; they can be revolutionary. Certainly this is true of the case at hand—the one that would require massive social change based on the simple proposition that it is wrong to kill other animals merely so that we might benefit. It this deceptively simple proposition is true, then *major* changes in how we live and what we live for, both as individuals and as a society, are morally required. And those who are threatened by these changes, those who ridicule the new ideas and the people who champion them—these reactionaries should be seen as real prisoners in a real cave: the cave of moral prejudice.

Living as if Life Matters

Suppose I am right about this—a large supposition, to be sure. If I am, then I can finally say something specific about the title of this essay—"Living as if Life Matters." The first thing I can say is rather simple. When we first think about what it means to live as if life matters we most likely understand this in terms of our own self-interest in living a full, rich life of our own—a point I return to in just a moment. What we are far less likely to think about is living a life in a way where *other* forms of life beyond the human species count for much.

Wise men and women who have gone before us have seen things differently. No less a figure than Albert Einstein speaks directly to this point. "A human being is a part of the whole, called by us the 'Universe,'" he writes, "a part limited in time and space. We experience ourself, our thoughts, and feelings as something separated from the rest—a kind of optical delusion of our consciousness. This delusion is a kind of prison for us, restricting us to our personal desires and to affection for a few persons nearest to us. Our task must be to free ourselves from this prison by widening our circle of compassion."[1]

"To free ourselves from this prison by widening our circle of compassion"—profound words, these, and profoundly relevant. If we draw our circle of the moral community in the way we have been taught, we will make some major ethical assumptions about what we are at liberty to do not only in our personal but also in our professional life. Medicine and science, various aspects of the legal profession and commerce, positions within the communications and advertising industries, even education—all these careers and many others can involve us in working to further the interests of companies or institutions that routinely kill other animals in the name of benefiting humans. Thus, if it really is wrong to kill other animals for this reason, then our present or future employment by these companies or institutions must be very difficult to defend morally. To those students within earshot, then, permit me to say: Before you decide on what career path to pursue, ask—*seriously* ask—whether the life that matters includes that of animals other than *Homo sapiens.* Whatever the right answer is, not to ask this question *seriously* is not to take truth seriously. It is that simple.

But there is another reason why all of us must ask this question, and ask it seriously. This concerns our own liberation from the cave. So long as we rest content with what we have been taught about right and wrong, we give sad testimony to the fact that our own life really does not matter much to us. A human being is not a robot, not a genetically hard-wired stimulus-response machine, not a body lacking a mind and will. Everything we know about human growth—about the flourishing of the human spirit and the actualization of those capacities that make a human life truly worth living—points to the need for each of us to take charge of our life, to nurture our rational autonomy, and to cultivate our capacity to care, not only about ourselves and others but also about what is true and just and good. Such growth can never occur if we remain happily enslaved to the unexamined moral conventions of our time and place and circumstances—our cave. Not only will we never glimpse the understanding that comes from knowing what makes right acts right, and what makes wrong acts wrong, we will likewise fail to understand *that* we do not understand and *why* we do not. Like the prisoners in the Allegory of the Cave we will mistake the shadows for the substance of truth and, in so doing, miss the opportunity to find our own true identity. And even if we never do find the answers we seek, the attempt to find them is itself liberating and ennobling. Here, surely, it is the journey not merely the destination that matters. And since that journey involves—and, indeed, must involve—doing moral philosophy, we have a second and, for now, a final way in which moral philosophy is relevant. Moral philosophy: both liberating and ennobling. Is there, I wonder, any branch of inquiry more important than this one?

3 *Ill-Gotten Gains*

Late in 1981 a reporter for a large metropolitan newspaper (we'll call her Karen to protect her interest in remaining anonymous) gained access to some previously classified government files. Using the Freedom of Information Act, Karen was investigating the federal government's funding of research into the short- and long-term effects of exposure to radioactive waste. It was with understandable surprise that, included in these files, she discovered the records of a series of experiments involving the induction and treatment of coronary thrombosis (heart attack). Conducted over a period of fifteen years by a renowned heart specialist (we'll call him Dr. Ventricle) and financed with federal funds, the experiments in all likelihood would have remained unknown to anyone outside Dr. Ventricle's sphere of power and influence had not Karen chanced upon them.

Karen's surprise soon gave way to shock and disbelief. In case after case she read of how Ventricle and his associates took otherwise healthy individuals, with no previous record of heart disease, and intentionally caused their heart to fail. The methods used to occasion the "attack" were a veritable shopping list of experimental techniques, from massive doses of stimulants (adrenaline was a favorite) to electrical damage of the coronary artery, which, in its weakened state, yielded the desired thrombosis. Members of Ventricle's team then set to work testing the efficacy of various drugs developed in the hope that they would help the heart withstand a second "attack." Dosages varied, and there were the usual control groups. Administering certain drugs to "patients" proved more efficacious in some cases than did administering no medication or smaller amounts of the same drugs in other cases. The research came to an abrupt end in the fall of

1981, but not because the project was judged unpromising or because someone raised a hue and cry about the ethics involved. Like so much else in the world at that time, Ventricle's project was a casualty of austere economic times. There simply wasn't enough federal money available to renew the grant application.

One would have to forsake all the instincts of a reporter to let the story end there. Karen persevered and, under false pretenses, secured an interview with Ventricle. When she revealed that she had gained access to the file, knew in detail the largely fruitless research conducted over fifteen years, and was incensed about his work, Ventricle was dumbfounded. But not because Karen had unearthed the file. And not even because it was filed where it was (a "clerical error," he assured her). What surprised Ventricle was that anyone would think there was a serious ethical question to be raised about what he had done. Karen's notes of their conversation include the following:

Ventricle: But I don't understand what you're getting at. Surely you know that heart disease is the leading cause of death. How can there be any ethical question about developing drugs which *literally* promise to be lifesaving?

Karen: Some people might agree that the goal—to save life—is a good, a noble end, and still question the means used to achieve it. Your "patients," after all, had no previous history of heart disease. *They* were healthy before you got your hands on them.

Ventricle: But medical progress simply isn't possible if we wait for people to get sick and then see what works. There are too many variables, too much beyond our control and comprehension, if we try to do our medical research in a clinical setting. The history of medicine shows how hopeless that approach is.

Karen: And I read, too, that upon completion of the experiment, assuming that the "patient" didn't die in the process—it says that those who survived were "sacrificed." You mean killed?

Ventricle: Yes, that's right. But always painlessly, always painlessly. And the body went immediately to the lab, where further tests were done. Nothing was wasted.

Karen: And it didn't bother you—I mean, you didn't ever ask yourself whether what you were doing was wrong? I mean . . .

Ventricle [interrupting]: My dear young lady, you make it seem as if I'm some kind of moral monster. I work for the benefit of humanity, and I have achieved some small success, I hope you will agree. Those who raise cries of wrongdoing about what I've done are well intentioned but misguided. After all, I use animals in my research—chimpanzees, to be more precise—not human beings.

The Point

The story about Karen and Dr. Ventricle is just that—a story, a small piece of fiction. There is no real Dr. Ventricle, no real Karen, and so on. But there *is* widespread use of animals in scientific research, including research like our imaginary Dr. Ventricle's. So the story, while its details are imaginary—while it is, let it be clear, a literary device, not a factual account—is a story with a point. Most people reading it would be morally outraged if there actually were a Dr. Ventricle who did coronary research of the sort described on otherwise healthy human beings. Considerably fewer would raise a morally quizzical eyebrow when informed of such research done on nonhuman animals, chimpanzees, or whatever. The story has a point, or so I hope, because, catching us off guard, it brings this difference home to us, gives it life in our experience, and, in doing so, reveals something about ourselves, something about our own constellation of values. If we think what Ventricle did would be wrong if done to human beings but all right if done to chimpanzees, then we must believe that there are different moral standards that apply to how we may treat the two—human beings and chimpanzees. But to acknowledge this difference, if acknowledge it we do, is only the beginning, not the end, of our moral thinking. We can meet the challenge to think well from the moral point of view only if we are able to cite a *morally relevant difference* between humans and chimpanzees, one that illuminates in a clear, coherent, and rationally defensible way why it would be wrong to use humans, but not chimpanzees, in research like Dr. Ventricle's.

The Larger Context

That we cannot rationally avoid this challenge is an idea that has only recently taken root in some quarters. Cora Diamond, a philosopher at the University of Virginia, notes that a recent bibliography on society, ethics, and the life sciences was described by its publishers as "containing the most pertinent references on precisely such subjects as experimentation, including, besides general material, sections specifically on experimentation on fetuses, prisoners, mental patients, and children"; the work includes *no* references to "the ethical problems of animal experimentation."[1] The explanation of this omission cannot be that there was, at that time, no (or not enough) literature on the topic, as even a cursory glance at Charles R. Magel's *A Bibliography on Animal Rights and Related Matters* will reveal.[2] By far the likelier explanation, as Diamond observes, is that "for many working in the field, the phrases 'ethical problems posed by research' and 'ethical problems posed by research on *human subjects*' are treated as simply interchangeable."[3] To treat them so is not to meet the challenge to give a clear, coherent, and rationally defensible basis for allowing research on other animals that we would not allow on humans. It is, instead, symptomatic of the moral prejudices of those who persist in assuming that there is no ethical challenge to be met.

Most people who attempt to meet this challenge tend to interpret it in biomedical terms. It is well to remind ourselves, however, of the magnitude and variety of animal use in scientific settings generally. Estimates of the total number of animals used for scientific purposes vary, some placing the total betwen twenty and forty million, others as high as one hundred million, just for the United States, just for a single year. Worldwide, the totals frequently given are more than twice these. Of these totals, perhaps about a fourth are used in medical research, given any uncontorted meaning of the expression "medical research" and allowing, as before, that estimates vary and are difficult to verify with anything approaching certainty. The remaining animals are used for instructional purposes (for example, in lab sections in standard biology classes in high schools and universities), in toxicity testing (in which "animal models" are used to estimate the risks and levels of harm humans are likely to run by using, or by being exposed to, the ever-increasing array of therapeutic and nontherapeutic products, from oven cleaners to eyeshadow, from asbestos to interferon), and in other scientific contexts.

We would serve our purposes ill, moreover, if we failed to remind ourselves of the variety of research that falls within the category of medical research as well as the multiplicity of means used to conduct it. Burn experiments (immersion of a part or the whole of an animal's body in boiling water, use of hot plates and blow torches, and, with research on internal burns, such as burns to the esophagus, lye, are among the methods employed); radiation research (dogs and primates are common "animal models" who are studied after exposure to both small and large levels of radioactivity, an ongoing type of medical research conducted in connection with weapons development by the military); drumming (animals are placed in a revolving drum whose internal protuberances break bones and bruise the flesh as a preliminary to the study of traumatic shock); brain research (cats and primates are favorite test animals, with drugs, electrodes, and surgical alterations, for example, used to influence and manipulate behavior)—these are a sample of the types of research and methods current in medical research. When the scope and intent of the research are more psychological than physical, the methods employed vary accordingly. Punishment experiments (these commonly involve electrical shock administered to the feet, tail, tooth pulp, or brain of, for example, cats, dogs, rabbits, primates, rats, or mice); immobilization research (for example, dogs are suspended in so-called Pavlovian slings or chimpanzees are strapped in restraining chairs); blinding and other investigations of sensory deprivation (for example, on cats in the course of studying sexual behaviors); aggression research (here test animals are induced by researchers to fight among themselves); stress experiments (any and all of the above methods, or loud noises, or random blasts of air can be used to produce stress, the effects of which may then be studied scientifically)—these (and there are many more) alternative approaches are illustrative of psychological research and methods involving animals.

No doubt some will deny the propriety of including some of the foregoing in the general category of medical research. These sorts of disagreements are to be expected. However they are resolved, the differences in the methods used in medical research, as well as the differences in the specific form such research takes, should not obscure their similarities. All such research, we may assume, has as its goals the advancement of human knowledge and the improvement of public health. These are laudatory ends. Our interest in what follows lies in morally assessing some of the means used to achieve them. Our own "moral research" will use Dr. Ventricle's work on chimpanzees as its "model."

The Law

Among the differences between chimpanzees and humans, one concerns their legal standing. It is against the law to do to human beings what Ventricle did to his chimpanzees. It is not against the law to do this to these animals. So, here we have a difference. But is it a morally relevant one?

The difference in the legal status of chimpanzees and humans would be morally relevant if we had good reason to believe that what is legal and what is moral go hand in glove: where we have the former, there we have the latter (and maybe vice versa too). But a moment's reflection shows how bad the fit between legality and morality sometimes is. A century and a half ago, the legal status of black people in the United States was similar to the legal status of a house, corn, a barn: They were property, other people's property, and could legally be bought and sold without regard to their personal interests. But the legality of the slave trade did not make it moral, any more than the law against drinking, during the era of that "great experiment" of Prohibition, made it immoral to drink. Sometimes, it is true, what the law declares illegal (for example, murder and rape) is immoral, and vice versa. But there is no necessary connection, no preestablished harmony between morality and the law. So, yes, the legal status of chimpanzees and humans differs; but that does not show that their moral status does. Their difference in legal status, in other words, is not a morally relevant difference and will not morally justify using these animals, but not humans, in Ventricle's research.

The "Right" Species

An obvious difference, one that is biological, not legal, is that chimpanzees and humans belong to different species. Once more, a difference certainly; but a morally relevant one? Suppose, for the sake of argument, that a difference in species membership *is* a morally relevant difference. If it is, and if A and B belong to two different species, then it is quite possible that killing or otherwise harming A is wrong, while doing the same things to B is not.

Let us test this idea by imagining that Steven Spielberg's E.T. and some of E.T.'s friends show up on Earth. Whatever else we may want to say of them, we do not want to say that they are members of our species, the species *Homo sapiens.* Now, if a difference in species is a morally relevant difference, we should be willing to say that it is *not* wrong to kill or otherwise harm E.T. and the other members of his biological species in sport hunting, for example, even though it *is* wrong to do this to members of our species for this reason. But no double standards are allowed. If *their* belonging to a different species makes it all right for us to kill or harm them, then *our* belonging to a different species than the one to which they belong will cancel the wrongness of their killing or harming us. "Sorry, chum," E.T.'s compatriots say, before taking aim at us or prior to inducing *our* heart attacks, "but you just don't belong to the right species." As for us, we cannot lodge a whine of a moral objection if species membership, besides being a biological difference, is a morally relevant one. Before we give our assent to this idea, therefore, we ought to consider whether, were we to come face to face with another powerful species of extraterrestrials, we would think it reasonable to try to move them by the force of moral argument and persuasion. If we do, we will reject the view that species differences, like other biological differences (e.g., race or sex), constitute a morally relevant difference of the kind we seek. But we will also need to remind ourselves that no double standards are allowed: Though chimpanzees and humans do differ in terms of the species to which each belongs, that difference by itself is not a morally relevant one. Ventricle could not, that is, defend his use of chimpanzees rather than humans in his research on the grounds that these animals belong to a different species from our own.

The Soul

Many people evidently believe that theological differences separate humans from other animals. God, they say, has given us immortal souls. Our earthly life is not our only life. Beyond the grave there is eternal life—for some, heaven, for others, hell. Other animals, alas, have no soul, in this view, and therefore have no life after death either. That, it might be claimed, is the morally relevant difference between them and us, and that is why, so it might be inferred, it would be wrong to use humans in Ventricle's research but not wrong to use chimpanzees.

Only three points will be urged against this position here. First, the theology just sketched (*very* crudely) is not the only one competing for our informed assent, and some of the others (most notably, religions from the East and those of many Native American peoples) do ascribe soul and an afterlife to animals. So before one could reasonably use this alleged theological difference between humans and other animals as a morally relevant difference, one would have to defend one's theological views against theological competitors. To explore these matters is well beyond the limited reach of this chapter. It is enough for our purposes to be mindful that there is much to explore.

Second, even assuming that humans have souls, while other animals lack them, there is no obvious logical connection between these "facts" and the judgment that it would be wrong to do some things to humans that it would not be wrong to do to chimpanzees. Having (or not having) a soul obviously makes a difference concerning the chances that one's soul will live on. If chimpanzees lack souls, their chances are nil. But why does that make it quite all right to use them *in this life* in Ventricle's research? And why does our having a soul, assuming we do, make it wrong *in this life* to use us? Many more questions are avoided than addressed by those who rely on a supposed "theological difference" between humans and other animals as their basis for judging how each may be treated.

But third, and finally, to make a particular theology the yardstick of what is permissible and, indeed, supported by public funds in a pluralistic society such as we find in twentieth-century America is itself morally objectionable, offending, minimally, the sound moral, not to mention legal, principle that church and state be kept separate. Even if it had been shown to be true, which it has not, that humans have souls and other animals do not, that should not be used as a weapon for making public policy. We will not, in short, find the morally relevant difference we seek if we look for it within the labyrinth of alternative theologies.

The Right to Consent

"Human beings can give or withhold their informed consent; animals cannot. That's the morally relevant difference." This argument is certainly mistaken on one count, and possibly mistaken on another. Concerning the latter point first, evidence steadily increases regarding the intellectual abilities of chimpanzees and other primates (e.g., gorillas). Much of the public's attention has been focused on reports of studies involving the alleged linguistic abilities of these animals, when instructed in such languages as American Sign Language for the deaf (ASL). Washoe. Lana. Nim Chimpski. Individual chimpanzees have attained international notoriety. But how much these animals do and can understand is very much up in the air at this point. Whether primates have sufficient ability to understand and use language and, if they do, whether they have sufficient ability to give or withhold their informed consent—these matters cannot be settled arbitrarily at this point in time. Possibly these animals lack these abilities. But possibly they do not. Those who trot out a doctrinaire position in this regard prove how little, not how much, they know.

Questions about the ability of chimpanzees to give informed consent aside, it should be obvious that this ability is not the morally relevant difference we are seeking. Suppose that, in addition to using chimpanzees, Ventricle also used some humans, but only mentally incompetent ones—those who, though they have discernable preferences, are too young or too old, too enfeebled or too confused, to give or withhold their informed consent. If the ability to give or withhold informed consent were the morally relevant difference we seek, we should be willing to say that it was not wrong for Ventricle to do his coronary research on these humans, though it would be wrong for him to do it on competent humans—those humans, in other words, who can give or withhold their informed consent.

But though one's willingness to consent to have someone do something to oneself may be, and frequently is, a good reason to absolve the other person of moral responsibility, one's inability to give or withhold informed consent is on a totally different moral footing. When Walter Reed's colleagues gave their informed consent to take part in the yellow fever experiments, those who exposed them to the potentially fatal bite of the fever parasite carried by mosquitoes were absolved of any moral responsibility for the risks the volunteers chose to run, and those who chose to run these risks, let us agree, acted above and beyond the normal call of duty—acted, as philosophers say, supererogatorily. Because they did more than duty strictly requires, in the hope and with the intention of benefiting others, these pioneers deserve our esteem and applause.

The case of human incompetents is radically different. Since these humans (e.g., young children and the mentally retarded) lack the requisite mental abilities to have duties in the first place, it is absurd to think of them as capable of acting supererogatorily; they cannot act "beyond the call" of duty, when, as is true in their case, they cannot understand that "call" to begin with. But though they cannot volunteer, in the way mentally competent humans can, they can be forced or coerced to do something against their will or contrary to their known preferences. Sometimes, no doubt, coercive intervention in their life is above moral reproach—indeed, is morally required, as when, for example, we force a young child to undergo a spinal tap to check for meningitis. But the range of cases in which we are morally permitted or obliged to use force or coercion on human incompetents in order to accomplish certain ends is not large by any means. Primarily it includes cases in which we act with the intention, and because we are motivated, *to forward the interests of that individual human being.* And that is not a license, not a blank check to force or coerce human incompetents to be put at risk of serious harm so that *others* might possibly be benefited by having *their* risks established or minimized. To treat the naturally occurring heart ailment of a human incompetent *is* morally imperative, and anything we learn as a result that is beneficial to others is not evil by any means. However, to intentionally bring about the heart attack of a human incompetent, on the chance that others might benefit, is morally out of bounds. Human incompetents do not exist as "medical resources" for the rest of us. Morally, Ventricle's research should be condemned if done on human incompetents, whatever benefits others might secure as a result. Imagine our gains to be as rich and real as you like. They would all be ill-gotten.

What is true in the case of human incompetents (those humans, once again, who, though they have known preferences, cannot give or withhold their informed consent) is true of chimpanzees (and other animals like them in the relevant respects, assuming, as we are, that chimpanzees cannot give or withhold their informed consent). Just as in the case of these humans, so also in the case of these animals, we are morally permitted and sometimes required to act in ways that coercively put them at risk of serious harm, against their known preferences, as when, for example, they are subjected to painful exploratory surgery. But the range of cases in which we are justified in using force or coercion on them is morally circumscribed. Primarily it is to promote *their* individual interests, as we perceive what is in their interests. It is *not* to promote the collective interests of *others,* including those of human beings. Chimpanzees are not our tasters, we are not their kings. To treat them in ways that put them at risk of significant harm on the chance that we might learn something useful, something that might benefit others (including other chimpanzees!), something that just might add to our understanding of disease or its treatment or prevention—coercively to put them at risk of significant harm for any or all of these reasons is morally to be condemned.

To attempt to avoid this finding in the case of these animals, while holding on to the companion finding in the case of incompetent humans, is as rational as trying to whistle without using your mouth. It can't be done. As certain as it is that it would have been wrong for Ventricle to use human incompetents in his coronary research, it is at least as certain that it would have been wrong for him to use chimpanzees instead, despite the legality of using these animals and the illegality of using these humans, and notwithstanding the actual biological and alleged theological differences between humans and chimpanzees. Whatever gains we might have harvested, for present or future generations of human beings, would have been ill-gotten.

The Indirect-Duty Response

People try to avoid this conclusion in a variety of ways. For example, some argue that we do not have any duties *to* animals (what philosophers call "direct duties"); rather, we have only duties *involving* animals (so-called "indirect duties"). Animals, in this view, have the same kind of moral status as redwoods, the Taj Mahal, and El Greco's *View of Toledo.* Few would deny that we have a duty to preserve these things, but most would deny that we have a duty *to them* to do so. Our duty, most people seem to think, is a duty *to other human beings,* both present and future generations, to preserve great works of art and the majesty of nature so that they, these other human beings, might have an opportunity to see and appreciate them, thereby enriching the quality of their lives. Duties involving works of art and the majesty of nature, in short, are indirect duties to humanity.

The same is true, some people maintain, of our duties regarding other animals. By all means, don't harm them unnecessarily, they say; but don't be misled into thinking that this is because we have duties directly to them. When animals are owned by others, we certainly ought not harm them unnecessarily because, after all, we have a duty to property owners not to harm their property. And when animals are not owned by anyone in particular, we still ought not harm them unnecessarily since people who do this to nonhuman animals have a tendency to do the same sorts of things to human beings; since we *do* have a duty not to do this sort of thing to human beings, we therefore ought to avoid doing it to other animals—not because we owe it to them, to be sure, but because we do owe it to one another.

If our duties regarding animals were indirect duties, one might then argue that Ventricle's research would have been wrong if done on human incompetents, but morally permissible if done on chimpanzees. If our duties regarding animals are indirect duties to humanity, then the morality of how we treat them is to be decided by what promotes human interests, and it is certainly possible that our interests would be promoted more by allowing animal research like Ventricle's than if we banned it.

But what about human incompetents? What type of duty do we have in their regard—direct or indirect? If one affirms direct duties in their case, while denying direct duties owed to chimpanzees, then, once again, one will want to be told what is the morally relevant difference between these humans and these animals, something that, as we know, has not been established by the arguments so far considered. Moreover, one cannot say that the morally relevant difference simply is that the duties owed to human incompetents are direct, while those involving chimpanzees are indirect, since this view presupposes that a morally relevant difference exists between the two. This view, therefore, cannot itself specify what that difference is.

The second option is to hold that our duties involving human incompetents, like our duties regarding chimpanzees, are indirect. This option at least has the merit of being consistent. Its principal defect is that it is false. Morally, it is preposterous to maintain that the reason why we ought not torture little children, for example, or kill their senile grandparents is because of the interests of others—for example, the children's parents or other elderly, more lucid people who, learning of the fate that befalls the senile, will live out their last years in wretched anxiety. It *is* wrong to torture children. But it is wrong to do this because in doing it we violate a duty we have directly to individual children *quite apart* from what their parents (or friends, strangers, etc.) happen to think or feel. And the same is true in the case of other harms we might visit upon other human incompetents. We owe it to them directly not to harm them. If others benefit in the bargain when we do as duty requires, they may count themselves lucky. But whether or not others benefit as a result of our refusal to harm human incompetents is, strictly speaking, morally irrelevant to whether we have duties to them. Our duties regarding human incompetents are not indirect duties to other people.

Of the two options, therefore, the second (that we have direct duties to human incompetents) is the one we should accept. Not to do so would be to distort, rather than illuminate, the moral status of these humans. Once more, however, we cannot consistently regard the moral status of chimpanzees any differently than that of these humans, if we are unable to cite a morally relevant difference. In other words, since our duties regarding human incompetents are direct duties, since we have duties regarding chimpanzees, and assuming we are unable to cite and defend a morally relevant difference between these animals and these humans, then our duties regarding chimpanzees are likewise direct, not indirect, duties—*duties we owe directly to them, considered as individuals.* In particular, therefore, we owe it to these animals themselves not to harm them. Any further moral thinking about these animals must both take this into account and be able to account for it, and any treatment of chimpanzees that rests on a view about these animals that is deficient in these respects cannot be rationally satisfactory. One cannot, therefore, defend the gains others might have received from Ventricle's research on chimpanzees, but condemn any gains stemming from such research if done on human incompetents, by claiming that our duties to these animals are indirect, while those involving these humans are direct. If one end is ill-gotten, then so is the other.

The Contractarian Response

Contractarianism is a second position that might seem to support research on chimpanzees but not on incompetent humans. Roughly speaking, contractarians view morality as consisting of a set of mutually agreed-upon constraints on everyone's behavior. Each party to these agreements (or "the contract"), we are to suppose, seeks to maximize what is in his or her individual self-interest. Each party soon realizes, however, that to achieve this objective, others must be limited in what they may do. For example, it is self-defeating for Friday to work to secure food and a place to live if Crusoe is at liberty to steal his property. Since no one has any self-interested reason to limit the pursuit of his or her self-interest unilaterally, such limits can come into being only if enough people agree to abide by them and, relatedly, agree to impose appropriate sanctions (e.g., fines or other punishments) on those who fail to cooperate.

Contractarians have important intramural differences. Some believe that present-day morality can be traced to an actual historical agreement ("the original contract"); others interpret the notion of a contract ahistorically. These and other internal differences to one side, it should be clear that nonhuman animals, chimpanzees included, can find a precarious home at best within standard versions of contractarianism. As far as we know, human beings are the only terrestrial creatures capable of entering into contracts. That being so, what duties, if any, we have regarding other animals must depend on what these human contractors judge to be in *their* (human) self-interest. If most of these humans agree that it is in their individual self-interest to allow Ventricle-like research on chimpanzees, while forbidding analogous research on human incompetents, then the former, but not the latter, research would be justified. In this way, then, contractarianism might seem to provide an adequate moral basis for Ventricle's research.

But contractarianism, at least given one of its expressions, could justify far more than research on nonhuman animals. If enough people happen to believe that it would be in their self-interest to suppress or oppress the members of a given minority (e.g., a racial or religious minority), when it comes to such vital matters as access to medical care, education, or career opportunities, then such policies, if mutually agreed upon by the majority, could not be morally condemned, given this version of contractarianism. This approach to morality, in other words, has the undesirable feature of legitimating the philosophy that might, understood as the collective judgment and power of the majority, makes right. Few, if any, will find this a congenial moral philosophy, since it would justify the most extreme expressions of racial and other forms of oppression. To have recourse to this philosophy as a defense of Ventricle-type research on chimpanzees, therefore, is like trying to keep one's moral position afloat by drilling a hole in it. Our (human) might does give us the power to use other animals in research, just as the might of the majority gives it the power to exploit the members of racial or religious minorities. But in neither case does might make right.

One could, it is true, endeavor to retain the spirit of contractarianism while altering the letter somewhat. Instead of allowing the parties to the contract to know, for example, their race, sex, religion, and nationality, one might ask them to imagine that they stand "behind a veil of ignorance," a veil that is thick enough to preclude their knowing the particular details of their life, thereby ensuring that they will select principles of justice impartially rather than on narrow, partisan grounds. Such a view can be found in John Rawls's influential work, *A Theory of Justice.* This is not the occasion to offer a full account or lengthy assessment of Rawls's version of contractarianism. Here it must suffice to note that Rawls, while insisting that the veil of ignorance keeps his contractors in the dark about, for example, what race or sex they will be, allows them to know that they will be members of the human race. Small wonder, then, that Rawls's view implies that we do have duties directly to one another but not to other animals. The cards, as dealt by Rawls, are stacked against ensuring an impartial judgment of the moral status of these animals.

Though his theory has much to recommend it, we will not find in it a rationally satisfying basis for defending research like Ventricle's.

The Utilitarian Response

A third view worthy of consideration concedes that chimpanzees are on all fours, so to speak, with human incompetents, when it comes to their respective moral status: Both are owed the same basic duty owed to those humans who are competent (that is, who have the ability to give or withhold their informed consent and, in having this ability, have all those other cognitive abilities thereby presupposed). That duty is twofold: first, to consider their interests, and second, to count equal interests equally. The interests in question are what individuals prefer or would rather do without—what they like or dislike, love or hate, what they are "for" or "against." In the case of every individual with interests, then, we must first take the time and trouble to ask what his or her interests are before we can decide what, morally speaking, we ought to do. More than this, however, we must also weigh or count equal interests equally. If Jack and Jill both want to go up the hill, and if they both want to do so equally, then we must count their interests in going up the hill as being equal in importance. Rationally, we cannot discount the importance of Jill's interest on the grounds that "she's only a girl" or Jack's because "he's a dumb jock." To treat Jack and Jill fairly, to treat them equitably, to treat them as equals, requires that we consider their respective interests and count their like interests as of like importance. Let us refer to the principle that demands equal consideration and weighting of like interests as the *equality of interests principle* or *equality principle.*

The equality of interests principle is one part of a currently fashionable view called "preference utilitarianism." Utilitarianism, very roughly speaking, requires that we act in order to bring about the optimum aggregate balance of good over bad consequences for all those affected by what we do. If we think of "the good" as involving the satisfaction of individual preferences, and "the bad" as involving the frustration of such preferences, then the close connection between the principles of utility and equality should be clear. As preference utilitarians, what we aim to bring about is the best aggregate balance of the satisfaction of preferences over their frustration for all affected by what we do. To aim at this objective, however, we must first consider who has what preferences (interests) and count equal interests equally; we must, that is, first rely on the equality of interests principle.

Some preference utilitarians think that preference utilitarianism would condemn Ventricle's research on chimpanzees. The arguments these thinkers give, however, are far from convincing. Essentially, what they come to is the claim that allowing research on the chimpanzees while forbidding it on human incompetents must violate the equality of interests principle, a principle that, given preference utilitarianism, it is always wrong to violate. But those who argue in this way are confused. Suppose both Jane Goodall and Figan (one of the chimpanzees she has studied in the wild) have the same interest in avoiding the excruciating pain associated with a heart attack. As preference utilitarians, we certainly must take the interests of both into account, and, assuming their equality, we must count them as being of equal importance. It does not follow from our having done this, however, that we must now approve of doing only those things to Jane Goodall that we would approve of doing to Figan, and vice versa. What we ought to do to either, assuming we have observed the strictures of the equality principle, is now to be determined by appeal to the principle of utility. And there is no reason why *the consequences for others* (namely, all those who will be affected by the consequences of our acts) will be the same if we do only the same things to Figan as we would be willing to do to Goodall. In particular, it is certainly possible that the aggregate balance of good over bad for all affected by the outcome would be better if Ventricle did his research on Figan than if he did it on Goodall.

It does not follow from this that preference utilitarians cannot condemn research such as Ventricle's when done on chimpanzees. What follows is, first, that they cannot condemn it on the ground that it must violate the equality of interests principle (for it need not), and, second, that they must acknowledge that whether or not they have grounds on which to condemn it depends on their having knowledge of the relevant consequences of Ventricle-like research. What consequences, then, are relevant? In the nature of the case, these must deal with the degree to which the interests (preferences) of all those affected by the outcome of the research are satisfied or frustrated. This requires more than our knowing how Figan's interests would fare. There are also Ventricle's interests to take into account, as well as those of his staff, plus those who build the tools of the medical researcher's trade (e.g., cages, restraint chairs), plus those who have an economic interest in the development of new drugs, plus our vital interest in health, and so on. There are, in a word, numerous interests to take into account and assess equitably before anyone could plausibly claim, with any degree of credibility, that the consequences of not allowing Ventricle to do his research on chimpanzees would not bring about the best aggregate balance of preference satisfaction over frustration for all those affected by the outcome. Indeed, given that what we are being asked to do is compare the interests of relatively few chimpanzees against the not unimportant (e.g., economic, scientific, and health) interests of many more human beings, the utilitarian case seems to bode ill for the chimpanzees. On the face of it, there is a very strong presumptive utilitarian case to be made in favor of Ventricle's research when done on chimpanzees.

This might seem to be good news for those who favor such research. It isn't. Preference utilitarianism does more than offer a way to justify Ventricle-type research when done on nonhuman animals. It will also justify similar research done on human incompetents or, for that matter, on human competents, even without or against their informed consent. Granted, in cases involving humans, just as in cases involving other animals, we must take pains to consider everyone's interests and count equal interests equally. Having done this, however, there is no reason why, in this or that case, the aggregate consequences for others might not be "the best" if we allowed research to be done on the humans in question. If we agree that research involving these humans is wrong, then we shall certainly want a moral principle that will not sanction it. That being so, we shall certainly want to avoid preference utilitarianism. It is no rational defense of Ventricle-type research on chimpanzees, therefore, to note that preference utilitarianism will, or very likely will, allow it. That view will, or very likely will, allow a great deal that is wrong. To show that preference utilitarianism would sanction research on chimpanzees is far from showing that such research is morally tolerable.

Perfectionist and Utilitarian Views of Value

Thus far we have advanced a controversial moral thesis—namely, that research like Ventricle's, when done on chimpanzees, cannot be justified by appealing to the benefits others do or might receive. And we have also considered, only to reject, various responses that seek to refute this thesis. Even if this thesis and its defense to this point are sound, much philosophical work remains to be done. A controversial thesis like the one before us does not stand on its own two feet. One must not only defend it against likely objections, a task to which we have attended, if incompletely, in the foregoing; one must also attempt to identify and defend the moral grounds on which the thesis stands, a task we have thus far failed to undertake. When, as now, we turn our attention to this item on our agenda, we must anticipate that the full weight of this challenge cannot be borne here.

Although preference utilitarianism is not the adequate position its advocates suppose, the emphasis it places on treating individuals as equals is an important corrective to less egalitarian visions of morality. The ancient Greek philosopher Aristotle, for example, offers a perfectionist moral theory: People are better (and so deserve more) than others if they have a certain cluster of intellectual and artistic excellences (virtues). Indeed, some people are so lacking in the favored virtues that Aristotle thinks they are born to be the slaves of those who are more generously endowed. Perfectionism of the Aristotelian sort must strike us as morally offensive, and it is one of the virtues of utilitarianism, because of the importance it places on treating relevantly similar individuals as equals, that it disassociates itself from perfectionism.

But the *type* of equality we find in preference (and other forms of) utilitarianism is easily misunderstood. For the preference utilitarian (to limit our attention to this version of utilitarianism), it is not individuals that count as equals but rather their mental states—their preference satisfactions and frustrations. Individuals are *receptacles* of value, things "into which," so to speak, what has value can be "poured," like liquid in a cup. But it is the liquid in the cup (that is, the mental states of satisfaction or frustration) that has value, whether positive or negative. The cup (that is, the individual human being or, for that matter, the individual chimpanzee), though "containing" what has value, has no value of its own.

To view humans or chimpanzees in this way is to offer a theory of their value (or, more precisely, their lack of it) that can legitimate using them as medical and other sorts of resources, when this theory of value is coupled with the utilitarian injunction to act in order to bring about the best aggregate balance of good over bad for all affected by the outcome. If we consider the interests of all those who will be affected, if we count equal interests equally, and if, having done this, we can bring about the best aggregate balance of good over bad for all affected by doing medical research on chimpanzees (or on human incompetents, or, indeed, on unwilling competent human beings), then our research is justified. The gains others secure, on this view, are *not* ill-gotten. To maintain, as we have, that such gains *are* ill-gotten is thus implicitly to reject preference utilitarian approaches to questions about the justification of medical research. More deeply and, for present purposes, more importantly, it is also to reject standard utilitarian theories of value. According to these theories, as noted, it is what "goes into" the individual, what the individual "contains"—for example, the mental state of satisfaction—that has value, not the individual. Our controversial thesis about using animals such as chimpanzees in research such as Ventricle's thus turns out to involve a different vision, neither perfectionist nor utilitarian, of the value of the individual.

The Value of the Individual

This alternative vision consists in viewing certain individuals as themselves having a distinctive kind of value, what we will call "inherent value." This kind of value is not the same as, is not reducible to, and is not commensurate either with such values as preference satisfaction or frustration (that is, mental states) or with such values as artistic or intellectual talents (that is, mental and other kinds of excellences or virtues). We cannot, that is, equate or reduce the inherent value of an individual to his or her mental states or virtues, and neither can we intelligibly compare the two. In this respect, the three kinds of value (mental states, virtues, and the inherent value of the individual) are like proverbial apples and oranges.

They are also like water and oil: They don't mix. It is not only that Jane Goodall's inherent value is not the same as, not reducible to, and not commensurate with *her* satisfaction, pleasures, intellectual and artistic skills, and so forth. In addition, *her* inherent value is not the same as, is not reducible to, and is not commensurate with the valuable mental states or talents of *other* individuals, whether taken singly or collectively. Moreover, and as a corollary of the preceding, the individual's inherent value is in all ways independent both of his or her usefulness relative to the interest of others and of how others feel about the individual (for example, whether one is liked or admired, despised, or merely tolerated). A prince and a pauper, a streetwalker and a nun, those who are loved and those who are forsaken, the genius and the retarded child, the artist and the philistine, the most generous philanthropist and the most unscrupulous used-car salesperson—all have inherent value, according to the view recommended here, and all have it equally. Decidedly nonperfectionist in letter and spirit, this vision of value is decidedly nonutilitarian as well.

What Difference Does It Make?

To view the value of individuals in this way is not an empty abstraction. To the question What difference does it make whether we view individuals as having equal inherent value, or, as utilitarians do, as lacking such value, or, as perfectionists do, as having such value but to varying degree?—our response must be, "It makes all the moral difference in the world!" Morally, we are *always* required to treat those who have inherent value in ways that display proper respect for their distinctive kind of value, and though we cannot on this occasion either articulate or defend the full range of obligations tied to this fundamental duty, we can note that we fail to show proper respect for those who have such value whenever we treat them as if they were mere receptacles of value or as if their value were dependent on, or reducible to, their possible utility relative to the interests of others. In particular, therefore, Ventricle would fail to act as duty requires—would, in other words, do what is morally wrong—if he conducted his coronary research on competent human beings, without their informed consent, on the grounds that this research just might lead to the development of drugs or surgical techniques that would benefit others. That would be to treat these human beings as mere receptacles or as mere medical resources for others, and though Ventricle might be able to do this and get away with it, and though others might benefit as a result, that would not alter the nature of the grievous wrong he would have done. And it would be wrong, not because (or only if) there were utilitarian considerations, or contractarian considerations, or perfectionist considerations against his doing his research on these human beings, but because it would mark a failure on his part to treat them with appropriate respect. To ascribe inherent value to competent human beings, then, provides us with the theoretical wherewithal to ground our moral case against using competent human beings, against their will, in research like Ventricle's.

Who Has Inherent Value?

If inherent value could nonarbitrarily be limited to competent humans, then we would have to look elsewhere to resolve the ethical issues involved in using other individuals (for example, chimpanzees) in medical research. But inherent value can only be limited to competent human beings by having recourse to one arbitrary maneuver or another. Once we recognize that we have direct duties to competent and incompetent humans as well as to animals such as chimpanzees; once we recognize the challenge to give a sound theoretical basis for these duties in the case of these humans and these other animals; once we recognize the failure of indirect duty, contractarian, and utilitarian theories of obligation; once we recognize that the inherent value of competent humans precludes using them as mere resources in such research; once we recognize that a perfectionist vision of morality, one that assigns degrees of inherent value on the basis of possession of favored virtues, is unacceptable because of its inegalitarian implications; and once we recognize that morality simply will not tolerate double standards, then we cannot, except arbitrarily, withhold ascribing inherent value, to an equal degree, to incompetent humans and other animals such as chimpanzees. All have this value, in short, and all have it equally. All considered, this is an essential part of the most adequate total vision of morality. Morally, none of those having inherent value may be used in Ventricle-like research (research that puts them at risk of significant harm in the name of securing benefits for others, whether those benefits are realized or not). And none may be used in such research because to do so is to treat them as if their value is somehow reducible to their possible utility relative to the interests of others, or as if their value is somehow reducible to their value as "receptacles." What contractarianism, utilitarianism, and the other "isms" discussed earlier will allow is not morally tolerable.

Hurting and Harming

The prohibition against research like Ventricle's, when conducted on animals such as chimpanzees, cannot be avoided by the use of anesthetics or other palliatives used to eliminate or reduce suffering. Other things being equal, to cause an animal to suffer is to harm that animal—is, that is, to diminish that individual animal's welfare. But these two notions—harming on the one hand and suffering on the other—differ in important ways. An individual's welfare can be diminished independently of causing her to suffer, as when, for example, a young woman is reduced to a "vegetable" by painlessly administering a debilitating drug to her while she sleeps. We mince words if we deny that harm has been done to her, though she suffers not. More generally, harms, understood as reductions in an individual's welfare, can take the form of either *inflictions* (gross physical suffering is the clearest example of a harm of this type) or *deprivations* (prolonged loss of physical freedom is a clear example of a harm of this kind). Not all harms hurt, in other words, just as not all hurts harm.

Viewed against the background of these ideas, an untimely death is seen to be the ultimate harm for both humans and animals such as chimpanzees, and it is the ultimate harm for both because it is their ultimate deprivation or loss—their loss of life itself. Let the means used to kill chimpanzees be as "humane" (a cruel word, this) as you like. That will not erase the harm that an untimely death is for these animals. True, the use of anesthetics and other "humane" steps lessens the wrong done to these animals, when they are "sacrificed" in Ventricle-type research. But a lesser wrong is not a right. To do research that culminates in the "sacrifice" of chimpanzees or that puts these and similar animals at risk of losing their life, in the hope that we might learn something that will benefit others, is morally to be condemned, however "humane" that research may be in other respects.

The Criterion of Inherent Value

It remains to be asked, before concluding, what underlies the possession of inherent value. Some are tempted by the idea that life itself is inherently valuable. This view would authorize attributing inherent value to chimpanzees, for example, and so might find favor with some people who oppose using these animals in research. But this view would also authorize attributing inherent value to anything and everything that is alive, including, for example, crabgrass, lice, bacteria, and cancer cells. It is exceedingly unclear, to put the point as mildly as possible, either that we have a duty to treat these things with respect or that any clear sense can be given to the idea that we do.

More plausible by far is the view that those individuals who have inherent value are *the subjects of a life*—are, that is, the experiencing subjects of a life that fares well or ill for them over time, those who have *an individual experiential welfare,* logically independent of their utility relative to the interests or welfare of others. Competent humans are subjects of a life in this sense. But so, too, are those incompetent humans who have concerned us. And so, too, and not unimportantly, are chimpanzees. Indeed, so too are the members of many other animals: cats and dogs, hogs and sheep, dolphins and wolves, horses and cattle. Where one draws the line between those animals who are, and those who are not, subjects of a life is certain to be controversial. Still, there is abundant reason to believe that the members of mammalian species of animals do have a psychophysical identity over time, do have an experiential life, do have an individual welfare. Common sense is on the side of viewing these animals in this way, and ordinary language is not strained in talking of them as individuals who have an experiential welfare. The behavior of these animals, moreover, is consistent with regarding them as subjects of a life, and evolutionary theory implies that there are many species of animals whose members are, like the members of the species *Homo sapiens,* experiencing subjects of a life of their own, with an individual welfare. On these grounds, then, we have very strong reason to believe, even if we lack conclusive proof, that these animals meet the subject-of-a-life criterion.

If, then, those who meet this criterion have inherent value, and have it equally, chimpanzees and other animals who are subjects of a life, not just human beings, have this value *and* have neither more nor less of it than we do. (To hold that they have less than we do is to land oneself in the inegalitarian swamp of perfectionism.) Moreover, if, as has been argued, having inherent value morally bars others from treating those who have it as mere receptacles or as mere resources for others, then any and all medical research like Ventricle's, done on these animals in the name of possibly benefiting others, stands morally condemned. And it is not only cases in which the benefits for others do not materialize that are condemnable; also to be condemned are cases, if indeed there are any, in which the benefits for others are genuine. In these cases, as in others, the ends do not justify the means. The *many millions* of mammalian animals used each year for scientific purposes, including medical research, bear mute, tragic testimony to the narrowness of our moral vision.

Conclusion

This condemnation of such research probably is at odds with the judgment that most people would make about this issue. If we had good reason to assume that the truth always lies with what most people think, then we could look approvingly on Ventricle-like research done on animals like chimpanzees in the name of benefits for others. But we have no good reason to believe that the truth is to be measured plausibly by majority opinion, and what we know of the history of prejudice and bigotry speaks powerfully, if painfully, against this view. Only the cumulative force of informed, fair, rigorous argument can decide where the truth lies, or most likely lies, when we examine a controversial moral question. Although openly acknowledging and, indeed, insisting on the limitations of the arguments in this chapter, these arguments make the case, in broad outline, against using animals such as chimpanzees in medical research such as Ventricle's. Various challenges to this position have been considered and judged inadequate, and the deeper philosophical grounds that stand beneath the surface of this controversy, grounds that concern alternative theories of the value of the individual (or lack of it), have been, if not thoroughly excavated, at least turned over. That does not bring thinking about value to an end, but it is something of a beginning.

Those who oppose the use of animals such as chimpanzees in research like Ventricle's and who accept the major themes advanced here, oppose it, then, not because they think that all such research is a waste of time and money, or because they think that it never leads to any benefits for others, or because they view those who do such research as, to use Ventricle's words, "moral monsters," or even because they love animals. Those of us who condemn such research do so because this research is not possible except at the grave moral price of failing to show proper respect for the value of the animals who are used. Since, whatever our gains, they are ill-gotten, we must bring an end to research like Ventricle's, whatever our losses. A fair measure of our moral integrity will be the extent of our resolve to work against allowing our scientific, economic, health, and other interests to serve as a reason for the wrongful exploitation of members of species of animals other than our own.

4 Why Child Pornography Is Wrong

Some acts that are wrong are paradigmatically wrong. By the expression "paradigmatically wrong" I mean any act that all rational, informed, impartial people of goodwill agree is wrong. The wanton torture of the innocent is an example of a paradigmatically wrong act. Any adequate moral theory must offer an intelligible account of why those acts that are paradigmatically wrong are wrong.[1]

Some acts involving children that may *be* wrong are not wrong paradigmatically. In some Baby Doe cases, for example, where a defective newborn is killed or allowed to die, it is possible that causing or allowing the child to die *is* wrong but not paradigmatically so. Rational, informed, impartial people of goodwill, that is, might make conflicting judgments about some of these cases, just as these same people might find themselves in disagreement over the ethics of suicide or euthanasia, for example. There are, one might say, limits to what we can know morally (assuming that we can know some things). When we enter the troubled waters of the nonparadigmatically wrong, tolerance may be a virtue greatly to be prized.

In contrast to the cases just mentioned, child pornography *is* an example of a paradigmatic wrong. Any adequate moral theory therefore must be able to account for it. As we shall see, few can.

The Face of the Victims

For purposes of this essay I shall use the interpretation of child pornography expounded in the 1982 U.S. Supreme Court decision in *New York v. Ferber.* This interpretation limits child pornography to works that appeal to the prurient interests of others by "*visually* depict[ing] sexual conduct by children below a specified age."[2] What this age is varies from state to state within the United States and from nation to nation. Some states recognize eighteen as the age of sexual consent, others specify sixteen. Among nations, some fix the age of sexual majority at eighteen, while Holland, for example, currently is considering lowering the age to twelve. Whatever the "specified age" (I shall assume it to be sixteen) the plain fact is that most of the victims are young, some as young as one week of age.

The kind of sexual activity filmed or photographed includes bestiality, masturbation, sexual intercourse of all varieties, oral sex, sadistic and masochistic displays, and lewd or lascivious exhibitions of the genitals or pubic area. The number of children involved is unclear. A 1977 estimate for Los Angeles placed the number for that city at approximately thirty thousand.[3] The national total must be in the hundreds of thousands. Worldwide, the numbers easily could be in the millions.

Runaways and homeless children likely are at greatest risk. But children from every class, race, family background, and religion may be and have been exploited. Most enter this subterranean world because of someone they know—a family member, teacher, friend, or neighbor, for example. As the *Report* of the Meese Commission on Pornography states, "many [of these children] are too young to know what has happened; others are powerless to refuse the demand of an authority figure; some seem to engage in the conduct 'voluntarily,' usually in order to obtain desperately needed adult affection."[4]

The *Report* indicates that comparatively little detailed research has been done on the immediate and lasting effects on children used in child pornography. The actual photographs or films frequently present the children as if they were enjoying themselves ("having a good time") but appearances can be deceiving. According to testimony presented before the Supreme Court in *Ferber,* "the use of children as subjects of pornographic materials is harmful to the psychological, emotional and mental health of the child."[5] Notwithstanding the paucity of careful research in the field, the *Report* declares that "the pain suffered by children used in pornography is often devastating, and always significant. In the short term the effects of such involvement include depression, suicidal thoughts, feelings of shame, guilt, alienation from family and peers, and massive acute anxiety."[6] In the case of those children old enough to understand what has been done to them, there is little reason to challenge the painful accuracy of this assessment.

The Face of the Criminals

Those who sexually abuse children, including those who use them for pornographic purposes, exhibit different psychopathologies and come from every social and economic strata. The Meese Commission distinguishes between "situational" and "preferential" molesters. "The former are people who act out of some serious sexual or psychological need, but choose children as victims only when they are readily and safely accessible. 'Preferential' molesters, on the other hand, are those with a clear sexual preference for children ('pedophiles' in common usage) who can only satisfy the demands of that preference through child victims. 'Preferential' abusers collect child pornography and/or erotica as a matter of course."[7] The *Report* does not estimate the number of each sort of molester, "but it does seem apparent," the Commission states, "that 'preferential' child molesters over the long term victimize far more children than do 'situational' abusers."[8] Certainly the former must be more responsible than the latter for production of and trade in child pornography.

Who are these child exploiters? Sadly, they come in all shapes and sizes, representing every economic, professional, religious, and racial background. There are cases of children being sold by their parents into the eager hands of the producers of child pornography, cases where the parents themselves are the criminals, cases where clergy and youth-group advisers (e.g., a local Boy Scout troop leader) have used their charges for pornographic purposes, and cases where—or so it has been alleged—preschool children, some as young as two years of age, have been sexually molested, abused, and filmed for pornographic purposes. Considered as a group, the victims of child pornography have no safe haven, just as, considered as a group, those who victimize them have no fixed identity: The criminal could be anyone.

The Kantian Account

A normative theory's adequacy, I have claimed, depends in part on its ability to account for the wrongness of what is paradigmatically wrong. Whether adult pornography is wrong is a much debated question. Even more debated is the question of what if any role the government should play in restricting the availability of material depicting adults in sexually explicit situations. The most commonly offered justification of government regulation of adult pornography—I note, parenthetically, that what is "most common" is not necessarily what is most reasonable—concerns the harms done to those who view it. Not so in the case of child pornography. The prevailing, sensible view is that sexual activity among children, coercively arranged for the purpose of child pornography, is wrong independently of any harm done to "the audience" (which is not to say that harm cannot be done to those who view it). It is the *children* who are wronged.

Everyone agrees, then—or so I must assume and hope—that child pornography is a paradigmatic wrong. Any adequate theory of obligation therefore must be able to account for (explain, illuminate) why it is wrong. The most influential normative theories shaping current thought in moral philosophy fail to pass this test. This certainly is true of Kant's normative theory. As is well known, his theory emphasizes the capacity of rational beings to be self-legislative. Consider the first formulation of the categorical imperative (the formula of universal law).

> Act so that the maxim of your action may be adopted as a universal law.

According to this formulation I am to act in such a way that the maxim of my action (the plan or intention that informs what I do) can without contradiction be imposed on every other rational being. Now, in some cases of paradigmatic wrongs (for example, a self-serving lie) the formula of universal law seems to offer a plausible account of why the act is wrong. But the situation is markedly different in the case of child pornography. For there is no "contradiction in the will" involved in universally legislating the use of children in pornography. Where a contradiction might arise is if a person were to will that every child except his or her own may be used for this purpose. But any "contradiction in the will" that might arise in this case can be erased if the person *is* prepared to will that her or his own child may be used for this end. And though we would (or in any event we should) look with grave disfavor on parents who were prepared to have their own children used for pornographic purposes, that fact—the fact that they are willing to do this—is not itself the reason or explanation of the wrongness of what they are willing to do.

Matters might appear more favorable for Kantians when we consider an alternative formulation of the categorical imperative: the formula of end in itself.

> Act so that you treat humanity, both in your own person, and in the person of every other, always as an end, never as a means merely.

Anyone who uses a child for pornographic purposes, it might be claimed, uses the child "merely as a means" and so violates this formulation of the categorical imperative.

This analysis expresses the spirit of Kant's theory but departs significantly from the letter. A *person* in Kant's theory—and recall that the current formulation prescribes how we are to treat humanity "in the *person*"—a person is a morally autonomous individual, the sort of being who is capable of acting in accordance with the first formulation of the categorical imperative. This means that a person must be able to formulate a maxim and then ask whether every other person is capable of acting on it. Possibly some children used in child pornography (for example, normal sixteen year olds) are capable of doing this. But not the two-year-old, preschool children mentioned earlier. And certainly not the one-week-old infant cited by the Meese Commission. That these children are *humans* is not to be denied. That they are *persons* in Kant's sense must be. The result is that the second formulation of the categorical imperative cannot even be directly applied to questions about how young children may be treated. And since that formula cannot be directly applied, whereas the formula of universal law would not prohibit the use of children for pornographic purposes, Kant's theory is unable to account for this great wrong.

The Act-Utilitarian Account

Act-utilitarians—those who, roughly speaking, require that we act to bring about the best balance of desirable over undesirable consequences for everyone affected by the outcome—act-utilitarians would seem to have a better chance at success than Kantians. Recall the claims endorsed by the Meese Commission: "The pain suffered by children used in pornography is often devastating, and always significant. In the short term the effects of such involvement include depression, suicidal thoughts, feelings of shame, guilt, alienation from family and peers, and massive acute anxiety." When such a variety of damaging short-term consequences routinely are experienced by the victims, one might insist that these harms will never be outweighed by the enjoyment and other benefits experienced by others.

But one's confidence in the adequacy of the act-utilitarian's account must be tempered by the recognition that the variety of damaging, short-term effects just listed may not always be present. In particular, since the effects in question (for example, feelings of shame, guilt, alienation from family and peers) are not within the standard psychological repertoire of very young children (say, those under two years of age), it is just false to claim that a victimized child of this age experiences these effects in the short term.

As for the long term, the situation is far from clear. We simply do not know what the lasting effects of isolated instances of sexual exploitation are for infants of a few weeks of age, for example. Doubtless the effects are highly detrimental for young children who are *routinely* exploited for pornographic purposes. Everything we know about normal child development points to this tragic conclusion. What is less clear is what the long-term developmental effects are likely to be when a very young child is used for these purposes, not often or routinely, but rarely or on only one occasion. Imagine a child of a week or a year who, for prurient purposes, is filmed during some sexually explicit activity (say, oral sex with an adult). Given that the damaging effects listed earlier cannot occur in the short run, and assuming that the child in question is subjected to such an experience on only one occasion, I think we must confess that we simply do not know whether the long-term effects on the child will be detrimental or, if detrimental, how detrimental.

This cannot be good news for act-utilitarians. On their account of the wrongness of child pornography, *everything* turns on our ability, in each case, to say *both* what the effects of child pornography are *and* how it is that the bad effects outweigh the good. If I am right and we sometimes simply do not know this, then act-utilitarians sometimes will lack the sort of information their theory requires *before we can say* what is wrong. But since we do know that child pornography *is* wrong, the act-utilitarian's account of its wrongness cannot be judged successful.

The problems child pornography poses for act-utilitarians actually are more severe than those sketched to this point. For while we sometimes are uncertain about the detrimental effects on the children, the pleasures secured by the producers and patrons of child pornography are painfully clear. Judged exclusively in terms of pleasure and pain (or satisfaction and dissatisfaction), it is not impossible that the amount of pleasure (or satisfaction) obtained by those with an interest in child pornography might exceed the amount of pain (or dissatisfaction) experienced by a child. And if this possibility should obtain, as well it may, then act-utilitarians would be obliged by their theory to say that child pornography *sometimes is not wrong.* Far from offering an adequate account of why child pornography is wrong, act-utilitarianism actually implies that it sometimes may be quite all right—permissible certainly, obligatory perhaps. On this basis, if on no other, we are right to disassociate ourselves from act-utilitarianism.

The Rule-Utilitarian Account

Rule-utilitarianism may seem to fare better. According to this version of utilitarianism we are to act on the basis of valid moral rules, the validity of the rules having been determined by the utility of having everyone abide by them. Why ought we to keep our promises and not kill the innocent? Because the world is a better place in which to live if everybody keeps their promises and does not kill those who are guiltless. Granted, there conceivably may be cases in which better results could be produced by killing an innocent human being (for example, in order to obtain an inheritance which one then puts to praiseworthy civic use—to build a children's hospital, say). Though possible, there is more to be lost than gained by allowing such discretion in observing the rule against homicide, according to the rule-utilitarian, which is why we ought to obey it.

Rule-utilitarianism thus seems to offer a way around the problems that are act-utilitarianism's undoing. For the rule-utilitarian the question we need to ask is whether better consequences will result if everyone always abides by certain rules concerning child pornography (for example: Do not produce child pornography, or support those who do so). To this question the rule-utilitarian will answer that it would be better if we had such rules and required universal compliance. True, there may be some cases where we do not know the short- or long-term effects on children used for pornographic purposes; and it is also true that the pain such a child experiences in a single episode might even be outweighed by the pleasures enjoyed by those who produce and seek out pornographic materials. But while either one or both of these truths is sufficient to deny any claim to adequacy made on behalf of act-utilitarianism, neither casts any doubt on the adequacy of rule-utilitarianism. Or so the rule-utilitarian might argue.

There are familiar objections to rule-utilitarianism. One asks whether it is a legitimate version of *utilitarianism.* It seems to demand that we not make exceptions to rules *even if doing so would lead to the best results in a given situation.* It is difficult to understand how someone can possibly be a utilitarian and hold this. Far more important for present purposes, however, is a second kind of objection. This concerns the method used by rule-utilitarians (and by any utilitarian for that matter) to determine what is right and wrong.

The method prescribed asks that we consider all the consequences of instituting a given rule. Everyone's interests must be taken into account, and equal interests must be counted equally. Whether a person is white or black, a man or a woman, a genius or a retarded child, that individual's interests must be considered. This is the heart of the expansive *egalitarianism* that utilitarians of every variety take great pride in. This is the utilitarian's fundamental way of *being fair* to everyone.

This conception of fairness must be operative when the rule-utilitarian is asked to decide whether to have a strict rule against child pornography. If everyone's interests must be taken into account, then the interests of the exploited child cannot be ignored. Nor can the interests of concerned parents. But neither can the interests of the producers and purveyors of child pornography. After all, these people *enjoy* abusing children for sexual purposes, they rather *like* seeing young bodies engaged in sexually explicit acts, many have a *real appetite,* a *real preference* for this sort of thing. These people would be terribly frustrated and disappointed if they were prevented from having their sexual desires satisfied. If we are to be fair to everyone involved, not just the children and their parents, then we must consider how the interests of the so-called pedophile (a lamentable corruption of what this word means), for example, will be affected if we enforce a strict prohibition against child pornography.

I think this state of affairs, which is created by applying the basic concept of fairness that is partially definitive of utilitarian theory, constitutes a *reductio* of this theory. It is morally grotesque, I think, to suppose that we must (if we are to be fair!) consider whether pedophiles will be disappointed or frustrated before we can fairly decide in favor of a strict prohibition against child pornography. The sexual appetites of these human beings have no role whatever to play in reaching this decision, and it is a symptom of a misguided theory that it should require otherwise. Do we seriously believe that the interests of sadistic torturers must be considered, in the name of being fair, before we can decide to prohibit recreational torture? Do we consult the interests of murderers before we prohibit murder? The questions answer themselves. The situation is no different in the case of the prohibition against child pornography. Some interests, preferences, or appetites *do not count and must not count* in our deliberations about what is right and wrong. Thus, since the egalitarian aspirations of utilitarianism, including any recognizable form of rule-utilitarianism, demand that everyone's interests, preferences, and the like be taken into account, whatever these interests, preferences, and the like may be, we are on sound theoretical and moral ground when we reject rule-utilitarianism. That we must have a rule that strictly forbids child pornography is true, but that we should validate this rule after the fashion of rule-utilitarianism is false.

Contractarianism

Variations on the theme of contractarianism also can be shown to be inadequate. Some forms of this theory allow the contractors to know who they are, what they like and want, what their aspirations and plans are. This much allowed, the contractors are then asked to decide what agreements they can strike with other rational agents, with the understanding that each person is to be motivated exclusively out of concern for his or her own self-interest. This version of contractarianism does not offer children much by way of protection.[9] Children, after all, especially those who are very young, do not have the wherewithal to know what is in their self-interest, let alone the capacity to enter into contracts. When it comes to their protection, everything depends on what self-interested adults agree to. And while one may profoundly hope that pedophiliac contractors will not meet with success in this quarter, it must be emphasized that what is right and wrong, given this approach to ethics, even including child pornography, is wholly contingent on what people decide is in their self-interest.

Thus, even if it happened to turn out that rational egoists at one time decided to prohibit child pornography, that agreement could be rescinded later on. For example, if the number of pedophiles grew large or powerful enough, so that rational egoists came to have a self-interested reason to allow child pornography, then this form of contractarianism would offer its seal of approval. I must assume that this implication offers more than enough reason to deny this theory's adequacy. Success in getting enough people to accept what is wrong does not make it right.

Other versions of contractarianism, the one favored by John

Rawls in particular, are more subtle and powerful.[10] Rawls asks us to imagine that we are ignorant of the details of our life. We do not know what we want, or where or when we live. Am I a man, a woman? A genius, a moron? An American, a Libyan? White, black? If I am to be fair, I must assume that I do not know these or any other details of my life. Only then can I make an unbiased selection of the principles of justice.

There is much to be said in support of Rawls's theory, and much to be said against it. How far the theory is supposed to reach, and whether in particular the theory is supposed to extend to rules about child pornography, is unclear. If the theory does not extend this far, then it is incomplete in a fundamentally damaging way: It fails even to offer an account of this serious paradigmatic wrong. On the other hand, if the theory is supposed to extend this far, then the situation is even worse. For the Rawlsian theory, if it does apply to child pornography, fails to offer an adequate account of why it is wrong.

At first glance it may seem otherwise. After all, each of us knows that, if we are born into the world at all, we will have a childhood. For this reason it may seem to be very much in our self-interest to choose rules or principles that will protect children, insisting on having a rule against child pornography in particular. However, for all I know, I will be a person who develops a real taste for child pornography in my later years. What, then, should I choose: (1) a rule that strictly prohibits child pornography, thereby trying to protect my interests while a child, or (2) not having such a rule, thereby taking my chances while a child while at the same time permitting the expression of my interests as a pedophile later in life, should I happen to become one? I do not understand how Rawlsian contractarianism can give a decisive answer to this question. That being so, I do not understand how this version of contractarianism can supply an adequate account of the wrongness of child pornography.

But Rawls's theory is worse than indecisive on this issue (assuming that it even can be applied to it). It fails more fundamentally because it permits us to count, *as being morally rele-*

vant, the interests of the child abuser. Since, for all I know, *I* may be one myself, should not I weigh my possible pedophiliac interests before I decide for or against having the rule? But these sorts of interests—self-interested interests—have *nothing* to contribute to the moral assessment of child pornography. In this respect Rawls's theory is just as inadequate, and inadequate for the same kind of reason, as standard versions of utilitarianism.

There are, of course, other familiar moral theories in addition to those discussed in the preceding paragraphs. Limitations of space prevent a close critical examination on this occasion. Here I must be content to say that those theories that have not been examined will suffer from one or another of the defects found in those that have been. Libertarianism, for example, is open to the objection that it fails to account for the strict prohibition against child pornography because it fails to offer a theory of rights that, if acted upon, *insures or guarantees* that young children, who are not yet morally autonomous, will be protected. And familiar interest theories of rights, which tie the possession of various rights to the presence of particular desires, also will come to grief for the same kind of reason. Young children are wronged by their sexual abusers even when these children do not have desires that are frustrated, thwarted, or otherwise denied by their adult exploiters.

I shall sketch part of a different sort of theory below, the one that I believe offers the best account of why child pornography is wrong. First, however, we need to take stock of what has been shown in the above.

Criteria of Adequacy

The preceding critique of familiar options in moral theory offers us some guidance regarding what will count as an adequate account of the wrongness of child pornography. From our examination of Kant we understand that the account we seek must (1) recognize and ground *direct* duties to children, even the very young. For its part the examination of utilitarianism reveals that (2) what makes child pornography wrong cannot be that the consequences are always bad, all considered (for they may not be in some cases). Finally, the shortcomings of Rawlsian and other forms of contractarianism show that (3) what makes child pornography wrong is not that it conflicts with what is in our own self-interest. The wrongness of child pornography, in short, must be accounted for independently of asking what is in my own interests, what the consequences are or will be for the child, and what the consequences are or will be for others (for example, the public at large). And this is to say that what makes child pornography wrong does not turn on considerations about future consequences. What is wrong with child pornography, what underlies its strict prohibition, must turn on *how the child is treated,* not on the consequences of this treatment.

Now, *how* children are treated in such circumstances (to use Kant's terminology) is as *mere means.* Someone with superior power (the adult exploiter) imposes his or her will on the child for purposes chosen by the exploiter, not the child. Indeed, in those cases where the child is very young the purposes are not even comprehended by the victim. Some children, of course, will in time grow into individuals who will understand such things and will be able to make autonomous choices. However, since a child's inability to understand coerced sexual activity, arranged for the prurient interests of others, does not cancel the wrong of using them for these purposes, it follows that the wrong that attaches to using them *as mere means* must turn on considerations about what the victims actually are, not on what the victims might become. So let us ask what the victims are.

The Species Reply

An obvious response is that the victims of child pornography are human children, and there is no doubt either that this is true or that most people are likely to rest content with this reply. And yet this answer demonstrably does not go far enough. What we want to know is ***what there is about being a human child*** that makes it wrong to use the child for pornographic purposes. We are not told what this is merely by being told that the child is human (a member of the species *Homo sapiens*). Species membership does not inform us of anything that is either morally relevant or decisive in judging what is right or wrong.

A Religious Reply

People with religious convictions have an answer to our question. Every human child is made in the image of God—is sacred. That is why using the child for pornographic purposes is the great wrong that it is. I do not think we should dismiss this response cavalierly. Unlike the simple appeal to species membership, religiously grounded responses to the question about what makes human life special have a presumptive relevance that we do well to consider.

It remains true, nonetheless, that religiously grounded conceptions of what human life is face very serious obstacles. First and foremost is the challenge to give reasons for believing in a particular deity since, without such reasons, there is no good reason to accept a particular answer to the question What is a human child? But, second, even if these obstacles could be overcome, the question would still remain, What is there about being a human being that makes humans so special in the eyes of God? To be told that we *are* special is not to be told *why* we are.

Some answers to this question are obviously deficient. It will not suffice to urge that humans are special because they are morally autonomous, are capable of entering into a loving relationship with God, or are called upon to represent the deity here on earth. This may be true of most human beings, but it certainly is not true of all (for example, the seriously retarded). And yet religious people *certainly* will not concede that it is therefore all right to use these humans for pornographic purposes.

In the face of these difficulties it is tempting to have recourse to claims about the immortal soul. That, it might be said, is why we humans are so special: We all have immortal souls. But not only is this another declaration of faith that, without the marshalling of considered beliefs in its behalf, stands just as unsupported as the declaration of faith in God; this new declaration fails to offer a credible account of *why it is wrong* to commit the most grievous of sins. Granted, if you murder me, then that will make a difference to what happens to you in the next life. Your terrible sin will be punished, let us assume, and that gives you a strong, self-interested reason for not murdering me. But *this* sort of reason (that it is not in your self-interest to murder me) fails altogether to capture why it is *morally wrong* to do so. What is wrong and what is contrary to one's self-interest are not two sides of the same coin. Declarations of belief in the immortality of the soul thus seem to be logically irrelevant to attempts to account for why such a heinous act as murder is wrong.

The same kind of argument can be given with respect to child pornography. Let us agree that this is a great wrong, a great sin. And let us concede that both the child and the child's exploiter have immortal souls. The question is, How does this latter claim help account for why child pornography is wrong? If the answer is given in terms of what will happen to child abusers in the afterlife, then what we have here is an appeal to the abuser's self-interest. And that sort of appeal, for reasons offered earlier, is not capable of accounting for the wrongness of child pornography. If, on the other hand, the harm done to the child is cited, then the situation is no better since appeals to a victim's immortal soul carries with it the promise that the (earthly) harm done to the child, though real enough, will be more than compensated for in the life beyond.

It would seem, then, that appeals to religious beliefs, including those that involve beliefs about the immortality of the human soul, are unlikely to provide us with the kind of reasoned account we are looking for when we ask why child pornography is wrong. And so it is that we are obliged to reconsider the question What *is* the human child, actually? from a different, less controversial, and hopefully more logically relevant vantage point.

The Respect Account

The view I wish to recommend and in part defend is the following. Children of whatever age are *subjects of a life.* By subject of a life I do not mean merely an individual who is conscious at one moment or another; rather, I mean an individual who has a psychophysical identity over time, an individual who has a biography and not merely a biology,[11] one who is an enduring somebody, not an enduring something—one who, to use more familiar terminology, is (or has) a "self." Those who read this essay are subjects of a life in the sense in which I am using this expression, and so are teenage victims of child abuse. But so, too, are very young children. True, some who are subjects of a life have the potential to grow into morally autonomous adults; and some subjects of a life actually are morally autonomous. But others lack this potential. However, even these individuals are biographical beings; at the most fundamental level, they are the same kind of being as their autonomous "cousins." The story of the psychophysical life of each is the story of a separate, distinct individual, each with a biography, each with an experiential life story of his or her own.

Here, then, are the makings of a different account of why child pornography is wrong. Its wrongness lies in treating a subject of a life as a mere means, as if that individual somebody (the abused child) were merely some thing. To express the same point in different words, the wrongness of child pornography consists in using coercion, deceit, or other reprehensible means to force one biographical being to satisfy the prurient needs, tastes, or preferences of another biographical being (or beings). Individuals who are stronger and more knowledgeable use their superior strength or knowledge to make weaker individuals serve as part of their (the stronger individuals') life stories. It is this coercive use of another biographical being that, in my view, underlies the wrongness of child pornography.

The normative position I am recommending can be expressed in other, more familiar terminology. The language of respect is the most obvious—(henceforth I shall refer to the account sketched here as "the respect account"). Using this language we can say that subjects of a life are always to be treated with respect, and this requires that we never treat them as if they exist for us, as our resources, as individuals who may be coercively, deceitfully, or in other ways treated merely as means to our ends. Those who traffic in child pornography do this to children, just as those who satisfy their sadistic desires through torture do this to their victims.

The respect account succeeds where the other accounts considered above fail. Recall the positive criteria of adequacy that emerged from the critique of these several options. Any adequate account of the wrongness of child pornography must (1) recognize and ground direct duties in the case of children and, without making reference to future consequences, must (2) recognize and ground the direct duty not to use children for pornographic purposes. The respect account satisfies both requirements. According to this account we have a *direct* duty, a duty owed *directly* to children, to treat them with respect, and one implication of this general duty is that we are not to use them for pornographic purposes. Moreover, according to the respect account, what is fundamentally wrong with abusing a child for pornographic purposes is *how the child is treated,* not what will happen to the child in the future (which is *not* to say that what happens in the future is never regrettable or even tragic). The child is wronged because the child is not treated with respect. And *that* is wrong *whatever* the future holds.

Some theories of obligation (for example, utilitarian theories) require that we consider the interests of those who exploit children for pornographic purposes before we can fairly decide whether this kind of exploitation is wrong. Any adequate account of why child pornography is wrong must deny the relevance of these interests. This is the third criterion of adequacy. The respect account satisfies it. The interests of different individuals are morally relevant only if acting on these interests does not conflict with treating subjects of a life with respect. The motivating interests of those who produce and use child pornography fail this test because these interests can be satisfied only at the cost of treating children as mere means, as mere resources, and not with the respect they are due. The respect account categorically denies the moral relevance of these interests, just as it similarly denies, for similar reasons, the moral relevance of the motivating interests of sadistic torturers. It is not "unfair" to exclude these interests or to refuse to count them "equally." What is unfair, and grievously so, is to allow these interests to play any role whatever in the determination of what is right and wrong.

Lastly, the respect account satisfies the criterion that the wrongness of child pornography not be made to turn on considerations about what is in our personal self-interest. The respect account satisfies this requirement because it locates the wrongness of child pornography in *what is done to the child, not whether and, if so, how this affects another's self-interest.* Even if child pornography were in our self-interest, that would not make it morally permissible; and even if it runs counter to our self-interest, that is not what makes it wrong. What makes it wrong has nothing whatever to do with anyone's self-interest. What makes it wrong is that children, who are subjects of a life, who are somebody, are treated as if they were some thing. The wrongness of child pornography consists in children's being treated with a lack of respect.

Conclusion

There are two final points I wish to make in conclusion. The first concerns a logical implication of my answer to the question Why is child pornography wrong? The second is more speculative.

Regarding the former point first, it is important to realize that the respect account does not limit the scope of respectful treatment to humans only. Religiously based ethics may do this (for example, on the grounds that only human beings have souls), and humanistic ethics certainly have this implication. But the respect account does not. This account calls for the respectful treatment of *all* subjects of a life, whether human or not. Given, then, that many nonhuman animals (mammals and birds, for example) *are* subjects of a life, the position stated and partially defended here demands that these animals are to be treated with respect. This we manifestly fail to do when, for example, we treat them as gustatory or scientific "resources" (when, that is, we raise and kill them for food, or use them as "tools" or "models" in research). According to the respect account, not only is our treatment of these animals wrong, it is wrong *for the same reasons* that child pornography is. Some may view this as an objection to this account; I regard it as one of its principal virtues. The day may come when we view the use of animals in science and their use as a food source (to mention only two of the more obvious forms of their ruthless exploitation) for the near-relatives of child pornography that they are.

Finally, if the analysis and argument of this essay are sound, then we ought to ask ourselves how it has been possible for most moral philosophers to fail to recognize the fundamental deficiencies of such influential theories as Kant's and Mill's, when challenged to account for the great wrong of child pornography. The obvious, chilling answer seems to be: Because moral philosophers have not devoted much attention to ethical questions about the treatment of children. But this answer only succeeds in forcing us to face another question, namely, Why have most moral philosophers failed to pay much attention to ethical questions about the treatment of children? My speculative answer to this question—and it is only speculative—is, Because most moral philosophers have been men, whereas the primary caretakers of our children have been women. Historically, the result of this division of labor has been that (male) moral philosophers have tended to ignore the moral ties that bind us to our children, a state of affairs that, like the role of women in the home, is changing. And none too soon. Informed attention to ethical questions involving children promises to improve both our homes and our moral philosophy.[12]

5 Feminism and Vivisection

Feminism takes more than one form and has more than one voice. I shall not be able to consider all of them. If I am sensible, this fact will chasten the confidence with which I reach my conclusions. Whatever form feminism takes, and with whatever voice the philosophy is spoken, we do best, I think, to ask how it responds to a recurring conception of the human person in Western moral and political thought—a conception that, while it hardly is lacking in critics, nevertheless is so pervasive that I think it appropriate to refer to it as "the traditional conception." This is the conception we find in such diverse thinkers as Thomas Hobbes, John Stuart Mill, John Locke, and Immanuel Kant, for example. Permit me to begin by sketching the main features of this widely shared view.

The Traditional Conception

The first point to note is the great emphasis these theorists place on *rationality,* not only its presence in normal humans, but also its value or importance. Hobbes's normative theory is a case in point. How are we to make our escape from the state of nature? Not by means of our brute force. Not by the pouring forth of feelings of sympathy, empathy, compassion, friendship, and love. No, given our selfish nature, it is only by means of our reason that we can set ourselves free of the "brutish" existence we otherwise are heir to. Mill, too, despite the many differences between his theory and Hobbes's, is enthusiastically on Hobbes's side when it comes to the place of reason in human life. It is reason, after all, that separates us from pigs, with whom we share an emotional kinship, and it is owing largely to our ability to take pleasure in the exercise of our rational capacities that we are able to forge a life that is really worth living—a life that cultivates the mind, not one that indulges the body. As for a natural-rights theorist such as John Locke, we are not surprised to learn that in his view only rational beings can have natural rights, and neither are we shocked when we discover that in Kant's theory the only individuals to whom we can have direct moral duties are individuals like us: rational beings. Indeed, in Kant's theory, our right acts have no moral worth at all if we perform them only because we want to do so; what worth they have is due exclusively to our exercising rational control, especially in the face of contrary desires. In all these theories, then, our identity as rational creatures is both affirmed and greatly valued, while at the same time our identity as physical beings, with roots in the biology of the past and entanglements with the ecology of the present, tend to be either entirely ignored or greatly minimized.

This celebration of reason over emotional and other noncognitive capacities can be purchased only at the price of denigrating the importance or value of these latter capacities. Feelings are not to be trusted; reason is. The emotions are to be—indeed, must be—controlled, and it is reason alone that is equal to the task. The carnal lusts of the body must be put in their place—must be tamed—and it is reason that cracks the whip. Present in all the theories, this denigration of our nonrational nature is perhaps most evident in Kant's, according to which reason is understood as a capacity of what he calls our "noumenal self," a power that transcends the natural world and operates freely of causal determinism. In these and other variations on this main theme we see that it is our connectedness with the rest of nature that is identified with that part of us that needs to be controlled or tamed. It is nature, expressing itself through our body, that requires a ruler, and it is what separates us from the rest of nature—our mind, our reason—that is assigned the task of trainer and controller. The wages of the body are sin. In reason alone lies salvation.

Along with this celebration of reason (the mind) over feelings or emotions (the body) we find a collateral celebration of the objective over the subjective. The standard of right conduct for me cannot be any different than the standard of right conduct for you, anymore than the sum of 2 plus 3 can differ between us. While Kant locates this objective standard in the test of universalizability (what he calls "the categorical imperative"), Mill offers the principle of utility; and whereas Hobbes fixes the standard of right conduct in those rules agreed upon by persons in the state of nature, Locke grounds morality in the universal "rights of Man." In every case, however, it is assumed that there is one and only one objective, impartial, abstract, universal basis of morality, the same for all people, at all times, and in all places. And since it is only reason, not our feelings or emotions, that can be counted upon to discover such a basis and, once having discovered it, apply it, it is dispassionate, objective reason, not passionate, subjective feelings that must guide us morally.

This theme of the desirability of domination by reason goes beyond the inner biographical struggles of individuals with themselves. It also gets played out when human nature is contrasted with nature generally. Nature is wild, treacherous, "nasty" (to use Hobbes's memorable description). Not only our own desires, but nature itself needs to be controlled, tamed, domesticated, and made to cooperate. Hobbes's acceptance of these views is perhaps the most obvious, but Locke, too, is a partisan: It is only *after* we humans have "mixed our labor" with some part of nature, only *after* we have taken it into ourselves as our "property," that it attains what value it has. For both Mill and Kant, moreover, the natural world, though it can be and sometimes is a source of aesthetic pleasure, awe, wonder, and sublimity, nevertheless lacks all these values apart from the appreciation of a human observer. Despite their many differences, therefore, all four of the major thinkers we have considered agree that nature has no value apart from human presence or use. It is we humans who give value to nature, especially through our use of our reason. It is mainly by the imposition of rational control over the natural world, done in the name of advancing our own and other human interests, and expressive of our achievements as cultural beings, that makes nature valuable.

Notice, further, how the spirit if not the letter of Hobbes's view, when interpreted as a historical thesis, remains alive in the other thinkers. It is, of course, false that humans lived the solitary existence Hobbes depicts. Everyone knows this. And yet each of the theories under review harkens back to that myth. We find this in Locke's emphasis on "the rights of Man"—of *individual* members of the species *Homo sapiens,* that is. And we also find it in Mill's celebrated "proof" of the principle of utility: It is, he argues, because *I* desire my own pleasure and view it as good, that I must desire that everyone have pleasure, and view that as very good indeed. And as for Kant, the noumenal self, which is the source of all freedom and worth in human life, *transcends* the physical world and, as such, can carry out its deliberations and make its decisions *as if it existed quite alone,* in a timeless vacuum. In sum, when all the dust settles, it is the interests, the pleasures, the freedom, or the rights of discrete, individual human beings, abstracted from their biological relationships to the natural world and their cultural relationships to the society in which they actually live, on which the fundamental principles of morality ultimately depend, according to the theories discussed in the above. And thus it is that, despite the demonstrable falsity of Hobbes's description of the state of nature, interpreted historically, these several moral philosophies continue to perpetuate the vision of individual human beings as "solitary" creatures—atoms of reason abstracted from their incidental biological and cultural molecular structure.

Just one final similarity should be noted. This concerns the normative role played by the idea of *impartiality.* To judge impartially is not to allow any personal factor to prejudice one's judgment. Judges would do this, for example, if they decided guilt or innocence on the basis of how well they knew those charged with breaking the law, how much they liked them, or what they stood to gain as a result of the verdicts they rendered. All four of the thinkers considered to this point are of one mind in disallowing partiality in the making of moral judgments—Locke in his affirmation of the basic rights of *all* humans; Mill in his brave demand that *everyone's* pleasures and pains must be considered and counted equitably; Hobbes in his insistence that the rules of morality be *fairly applied to everyone;* and Kant in his categorical requirement that *every one of us* must ask whether our individual maxims can be universalized, whether, that is, *everyone* could act as we do, for the same reasons we have. In a word, morality for all four thinkers requires the defeat of partiality and the triumph of impartiality.

Here, then, we have, by way of summary, the recurring image of the human person—the individual human being—that has characterized so much of Western moral and political thought. Each of us is by nature selfish, innately predisposed to look out for only our own good. Moreover, this deeply rooted selfishness is expressed by our standing apart from, rather than in our being a part of, a biological, ecological, or social community. Indeed, the most fundamental task we all face is how to meld our natural appetite for our individual good into the larger fabric of life, especially human society. The solution to this problem is not to be found in our emotions or feelings, which are irrational and liable to mislead us and which in any event we share with many other animals. Instead, it is to be found in our ability to reason, which is unique to the human species. Once this is understood, the key to unlocking the essence of a good hunan life also is apparent. This is to be found in cultivating that which is uniquely human—our reason or, more generally, our mind. Only in this way can we overcome the limitations placed on our fulfillment by the body and its demands. This triumph of the mind over the body also carries over into our dealings with the natural world, which may be viewed as a kind of shared body, standing no less in need of being tamed and domesticated than our individual physicality. The successful domestication of nature, in which we may all take part, is what is known as culture. The greater the triumph over nature, the higher the culture, the less so, the more "primitive" the culture and the people comprising it. In the particular case of a fully developed moral life, finally, growth lies in the direction of objectivity as opposed to subjectivity, impartiality as opposed to partiality, and—of course—reason as opposed to emotion.

Liberal Feminism

Now, some feminist thinkers are not critical of the views I have just summarized. For these thinkers the problem does not lie in these views (not in the traditional conception) but in how women have been denied those very things that are essential to a good, full human life, given this conception of the human person. Mary Wollstonecraft speaks admirably for this wing of feminist thinkers, arguing in her classic 1792 work, *A Vindication of the Rights of Woman,* that women are no more lacking in the uniquely human characteristics than are men.

> In what does man's [that is, humanity's] pre-eminence over the brute creation, consist? The answer is as clear as that a half is less than a whole, in Reason. . . . For what purpose were the passions implanted? That man by struggling with them might attain a degree of knowledge denied of the brutes.
>
> Consequently the perfection of our nature and capability of happiness must be estimated by the degree of reason, virtue and humanity that distinguish the individual and that from the exercise of reason, knowledge and virtue naturally flow.[1]

The mention of "virtues" should not be misunderstood. Commenting on Wollstonecraft's thought, the Australian philosopher Val Plumwood remarks that "*masculine* virtues are taken to be *human* virtues, what distinguishes humans from the sphere of nature, especially the qualities of rationality, transcendence and freedom. . . . The human character ideal she [Wollstonecraft] espouses diverges sharply from the feminine character ideal, which she rejects, 'despising that weak elegancy of mind, exquisite sensibility, and sweet docility of manners.' Instead, she urges that women become 'more masculine and respectable.' "[2]

That "masculinity" would find expression in power and domination, a point not lost on Amelia Bloomer who, writing in the nineteenth century, asserts that "He [man] has brought the Bible to prove that he is her [woman's] lord and master, and taught her that resistance to his authority is to resist God's will. I deny that the Bible teaches any such doctrine," Bloomer asserts, concluding "God made them different in sex, but equal in intellect, and gave them equal dominion."[3] When it comes to domination, it seems, two can play as well as one.

As these representative quotations from Wollstonecraft and Bloomer show, some feminist thinkers are prepared to work with the traditional concept of the human person rather than to challenge it. What *they* challenge are those social and legal impediments that stand in the way of *equality of treatment* of both women and men, a concept of justice at the very heart of the liberal tradition in Western political thought, which helps to explain the appropriateness of the name—"liberal feminism"—now commonly applied to this strain of feminist thought.

To a limited degree liberal feminists have succeeded in achieving some of their goals. Educational and career opportunities, for example, formerly closed to women, increasingly are open to them. Some thinkers (sometimes called "sexual conservatives") are of the opinion that liberal feminists already have gone too far. But other thinkers within the feminist movement in philosophy believe that liberal feminists have not gone far enough. Indeed, the very things liberal feminists would point to as part of the solution—things like greater educational and professional opportunities for women, for example—in the eyes of some feminists actually are seen as part of feminism's major problem.

The Feminist Critique of Liberal Feminism

Those feminists who are critical of the liberal position are a mixed group, with no univocal voice or identity. I will not be able to discuss all of them. But allow me briefly to sketch a common line of criticism they raise against liberal feminism.

According to these critics, it is not enough that women be able to compete equitably with men for educational opportunities, jobs, pay, or social and professional benefits. All that these changes would achieve, were they to occur, is what Ariel Salleh has described as the "masculinising" of feminism.[4] Val Plumwood explains this idea when she writes that "the problem for women [according to liberal feminism] was to claim full *humanity* i.e. conform to the main human character ideal, defined by traits characteristic also of the masculine, and to fit into, adapt themselves to, the corresponding social institutions of the public sphere. These might require some minor modification but basically it was women who were to change and adapt, (sometimes with help) and women (or what society had made of them) who were the problem. The position can be summed up as that of demanding equal participation in a masculine concept or ideal of humanity, and the associated activist strategy as that of demanding equal admittance for women to a masculine-defined sphere and masculine institutions."[5]

Anyone reading this will have noticed how heavy-laden with the word "masculine" this passage from Plumwood is—"masculine institutions," "masculine-defined sphere," "masculine concept or ideal of humanity." Why does she express herself in these terms? One way of answering this question, I think, is as follows.

Historically, in Western civilization at least, it has been men—not all men of course, but, in comparison with women, a greatly disproportionate number of them—who have been afforded opportunities to receive an education, thereby cultivating their minds; it has been men who have been given the freedom to "mix their labor" with nature and thereby come to be property owners; it has been men who have been encouraged to pursue their interests outside the home, wherever they may lead; and it has been men who have been in a position to make major contributions to our culture. Women, by contrast, have been assigned the roles of satisfying men's sexual needs (as "sex objects"), and of running the home (without pay) and raising the children (again without pay), with the result that, comparatively speaking, few have been able to acquire property, *or* pursue their interests beyond the home, *or* secure an education or job that was comparable to a man's, *or* contribute in large numbers to the forging of our culture—to our art, commerce, religion, technology, law, literature, and (alas) our philosophy.

Viewed against the backdrop of the ideas presented earlier, what we can see, then, is that men, to a vastly greater extent than women, have been given opportunities to realize the sort of good, full human life that emanates from acceptance of the traditional conception of the human person. The interpretation of this fact, offered by such thinkers as Ariel Salleh, is that this traditional conception of the human individual and the allied one of a good, full human life are not the sexually unbiased, objective, impartial views they might at first appear to be; on the contrary, they are sexually biased, discriminatory, ideological views of what *men* think the human person is or can be, and of what a good, full human life is, *if you are male.* Only instead of these normative beliefs being put forward in this plain, frank language, they are expressed in the language of *humanity,* embodying an ideal of a good, full human life *for all human beings.* The message is clear: Men *are* the measure of all things—or, at least, of *these* things.

The upshot is that the kind of life that historically has been available to most women has been viewed as vastly less than a good, full human life. And the virtues traditionally associated with womankind—the virtues associated with nurturing children and sustaining a home, those identified with empathy, sympathy, compassion, gentleness, and the "softer" emotional side of human nature—these virtues normally have been relegated to a very low standing by men, as virtues no *man* would (or should) value in another man or desire in himself. Small wonder, then, that thinkers like Wollstonecraft, herself enraptured by the traditional conception of the human person, would abjure the "feminine" virtues and encourage her sisters to become "more masculine and respectable." Once the traditional conception is accepted, the only way women have to achieve a truly worthwhile life is to emulate the ways and virtues traditionally open to men—for example, cultivating their reason (as opposed to their "exquisite sensibility") and claiming their fair share of the right to dominate nature by owning parts of it.

Thus are we able to understand why Plumwood expresses herself in the way she does (recall her use of such expressions as "masculine institutions" and "masculine concept or ideal of humanity," for example), and why feminists like Salleh wish to part company with the liberal feminist position. In the simplest possible terms, Salleh and the other feminists she represents believe that liberal feminism is not a true feminism but is, rather, a "masculinism"—a case of women trying to adjust themselves to male categories, male values, male institutions, and to the expectations of (to use a word much favored in feminist literature) patriarchy. At the risk of oversimplification, but in the interests of simplicity, I shall refer to this wing of feminism as *ethic-of-care feminism.*

Ethic-of-Care Feminism

One way to approach this wing of feminism is to begin with some observations of Marilyn French. French sees in the liberal feminist's acceptance of the traditional conception of the human person the regrettable acceptance of patriarchy. In her view a true feminism must reject patriarchy, not adjust itself to it. She writes as follows:

> Patriarchy is an ideology founded on the assumption that man is distinct from the animal and superior to it. The basis of this

> superiority is man's contact with a higher power/knowledge called god, reason, or control. The reason for man's existence is to shed all animal residue and realize fully his "divine" nature, the part that *seems* unlike any part owned by animals—mind, spirit, or control. In the process of achieving this, man has attempted to subdue nature both outside and inside himself; he has created a substitute environment in which he appears to be no longer dependent upon nature. The aim of the most influential minds has been to create an entirely factitious world, a world dominated by man, the one creature in control of his own destiny. This world, if complete, would be *entirely* in man's control . . . , and man himself would have eradicated or concealed his basic bodily and emotional bonds to nature.[6]

When French says that patriarchy as an ideology is *founded* on the assumptions of man's distinctness from and superiority to "the animals," she means, I think, that the other features of a patriarchal view of the world—for example, the celebration of reason over emotion, and of the mind over the body—are natural (though not necessarily logical) outgrowths of this more basic assumption. Thus, to attack patriarchy at its root would be to attack it at this point—the point where it is assumed that *Man* is "distinct from the animal and superior to it." And this in turn is to say that the object of critical scrutiny is and must be the traditional conception of the human person.

Two very different lines of criticism can be pursued here. The first insists that women, no less than men, are "distinct from the animal and superior to it." This is the line of criticism we find in Mary Wollstonecraft and other feminists who argue for the equality of women with men and who call for fairer, more equitable treatment for women. In other words, this is the familiar trail blazed by liberal feminists, a trail that in Ariel Salleh's telling phrase can only lead to the "masculinizing" of feminism, if feminism stops there. For Ariel Salleh and other feminists, however, a true feminism cannot stop there. A true feminism must dismantle the ideology of patriarchy, not only at its branches, so to speak—not only, that is, in those laws and social practices that discriminate unfairly against women—but must dismantle it *root and all.* And this means—or so I believe—that a true feminism must disassociate itself from the basic assumptions of patriarchy, including the one identified by Marilyn French: that "man is distinct from the animal and superior to it."

The Ethic of Care

Now, the beginning of this disassociation lies in our acceptance of our *animality.* Social by nature we may be (as Aristotle observes and, on one interpretation, Hobbes contests), but animal we most certainly are. This is not to demean ourselves, not to make ourselves something less than what we are. Or, rather, the very thought that, in acknowledging and accepting our animality, we are demeaning or lessening ourselves—this *very thought* is itself a symptom of the ideology that needs uprooting.

But now there is a problem. Indeed, now there are quite a few problems. I shall explore only two. To begin with, if the abandonment of patriarchy really does involve not only acceptance of our animality but also rejection of our superiority over all other animals who lack a soul, or reason, or control, then we presumably must come to acknowledge our *moral equality* with those animals who are otherwise like us—those animals who, that is, are conscious of the world, not only present in it, and who, like us, respond to the world emotionally (through fear and anxiety, for example) and who can be pleased or pained by what happens to them. Thus, if an enlightened feminist perspective would find it wrong to make a human animal suffer, in a given context, or to cause a human fear or anxiety, it seems that this same perspective must make the same finding if a nonhuman animal is made to suffer, or is caused fear or anxiety, in a similar context. As I say, this *seems* to be the case. The first problem I want to address is how, viewed from a feminist perspective, this can be the case.

The moral status of nonhuman animals poses at least two problems for many feminist philosophers. First there is the notable influence exercised by Carol Gilligan and her feminist moral theory—her "ethic of care," which she distinguishes from the patriarchal moral traditions of individual rights and duties.[7] There is much about Gilligan's views and the debates they have engendered that I am unable to recount and assess on this occasion. Suffice it to say here that if we accept the ethic of care as our ruling paradigm, we will naturally emphasize the importance of nurturing and sustaining existing interpersonal relationships (for example, those between parent and child, friend and friend, lover and lover). We will try to avoid conflicts and, when they do arise, we will try to find mutually beneficial ways to resolve them.

This is all well and good, as far as it goes. But does it go far enough? In particular, what are the resources within the ethic of care that can move people to consider the ethics of their dealings with individuals who *stand outside* the existing circle of their valued interpersonal relationships? Clearly, the resources seem to depend entirely on *how much one cares* for these individuals. And yet this seems to be such a contingent, such a "chancy" basis for such an important moral idea. The very bonds of caring that unite us with some, divide us from others. It is certainly possible that we should care about what happens to some individuals when in fact we do not care at all. And unless we supplement the ethic of care with some other motivating force—some other grounding of our moral judgment—we run the grave risk that our ethic will be excessively conservative and will blind us to those obligations we have to people for whom we are indifferent.

Nowhere, perhaps, is this possibility more evident than in the case of our moral dealings with nonhuman animals. The plain fact is, most people do not care very much about what happens to them, or, if they do care, their care seems to be highly selective, limited to "pet" animals, or to cuddly or rare specimens of wildlife, for example. What, then, becomes of the animals toward whom people are indifferent, given the ethic of care? Notice that we cannot say—what I find it natural to say—that we *ought* to care for these animals, that we have a *duty* to care for them, and, indeed, that they have a *right* to our protection. To speak in these natural ways—in the language of individual rights and duties—is out of bounds, given the ethic of care, and it is out of bounds because it is supposed to be a veiled expression of patriarchy.

But if these ways of speaking are ruled out, and if the moral ties that bind us to others are those defined by the ethic of care, then it seems doubtful that we will have the resources within our theory to be able to move people from their current attitudes of indifference regarding nonhuman animals to new, more caring ones. And thus it is that a feminist ethic that is *limited to an ethic of care* will, I think, be unable to illuminate the moral significance of the idea that we (human) animals are not superior to all other animals—and that, indeed, we and they are moral equals. For where the care is unequal, and the vocabulary of duties and rights has no voice, one's ethical options seem to be exhausted. This cannot be a salutary situation for the ethic of care.

A Place for Reason in Ethics

That is the first problem the moral status of nonhuman animals poses for ethic-of-care feminism. The second problem arises when we stop to consider ways in which the circle of care might be extended to include these animals. One way—though perhaps not the only, or even the best, way—in which this might be done is by having recourse to what I call "consistency arguments." The point of such arguments is to highlight the moral arbitrariness of excluding some individuals from the circle of one's moral concern. For example, pain is pain, wheresoever it may be, and the untimely death of anyone, especially the innocent, prima facie makes the world less good than it otherwise would have been. Granted, we care about pain and death most when those who suffer or die are close to us, and perhaps some people care even more when it is their own suffering and death they must confront. And yet this caring about the pain and death of another would not make sense if pain and untimely death themselves were unimportant or undesirable. These great evils are prima facie undesirable apart from the particular identities of those who suffer and those who die.

This much granted, reason is now empowered to demonstrate the arbitrariness of the limits imposed by an ethic of care. Suppose it is true that we do not care much about the pain and death of strangers—for example, the pain and death of the many homeless people eking out a life (an *existence* would be closer to the truth) on the streets of major urban areas. Our presumed indifference toward them is not a sufficient reason to deny that we have any obligations to them. Since pain and an untimely death are undesirable apart from the particular identities of those who suffer or die, they are no less undesirable if those who suffer and die are strangers than if they are friends. Our obligation to help others, therefore—for example, to prevent their avoidable suffering—holds whether we care about or are indifferent to the one who needs our assistance. Perhaps it is true that we are more likely to succeed in helping those people (friends or our children, for example) we know best, and perhaps it is also true that, in many cases, our most stringent obligations are to those with whom we have a valued interpersonal relationship. But this cannot be the whole of morality. The moral universe is not circumscribed by the caring universe. In many cases reason shows that we should help another, even when we do not care for that individual.

This same kind of argument (a consistency argument) also can be marshalled in the case of the moral status of nonhuman animals. For if pain and an untimely death are undesirable in the case of humans, whatever their identities, then the same must be true of relevantly similar nonhuman animals, whatever their identities. In this way, then, an ethic of care can seek to overcome its conservative tendencies and to illuminate how nonhuman animals can gain entry to the moral community, as beings who are to be protected by some of the same principles that are to protect human beings, even if we do not care about what happens to them.

Moreover, advocates of an ethic of care can attempt to achieve these normative results without renouncing the importance of care in a fully developed moral life. Indeed, one might even urge (as I am inclined to do) that one of the great challenges the ethic of care places before us is to bring our caring into line with our reason, in the sense that in time we *do* care about the pain and death of those for whom reason informs us that we *should* care.

Now, this way of addressing the first problem sketched above—the problem feminists face if they attempt to challenge the doctrine of human superiority while relying on the ethic of care—creates a second problem for these thinkers. For if the most (or at least a) satisfactory way to overcome the moral conservatism of the ethic of care, especially as this regards the assault on human chauvinism, is to make use of consistency arguments, then the feminist critique of patriarchy may not be as radical as some might wish it to be. For since these arguments by their very nature involve appeals to reason and abstract principles of logic, and since such principles and appeals are by their very nature partially definitive of the traditional conception of the human person (the very conception judged to be patriarchal to the core), it emerges that ethic-of-care feminists themselves must traffic with identifiable elements of the very prejudice they seek to overthrow.

To put the point more simply, these feminists face an unavoidable dilemma: *Either* their ethical theory lacks the theoretical wherewithal to mount the desired attack on human chauvinism, *or* it contains the means to mount this attack. If the former, then the theory is deficient because of the barrenness of claims to human superiority. If the latter, then the theory is still deficient, judged by its own standards, since in this case it must incorporate elements of the very conception (the patriarchal conception) of the human person that it seeks to replace. Of the two alternatives, it is the latter one that is the least damaging and the one that, therefore, I believe ethic-of-care feminists should accept. But whichever option is chosen, ethic-of-care feminism cannot be as radical as some might wish it to be.

Ethic-of-Care Feminism and Animal Liberation

Suppose I am right, and ethic-of-care feminists are at liberty to use consistency arguments to wage their attack on the sort of human chauvinism that is indigenous to a patriarchal understanding of the world. If this much is granted, then we are in a position to ask what views these feminists should have concerning human interactions with nonhuman animals, if not in general then in some more or less clearly defined context—say, the use humans make of nonhuman animals in science. To answer this question, we first need to offer a general characterization of how these animals are used in science.

The answer to this question is simple and straightforward. Whether the scientific interest be that of basic or applied research, whether it be product testing or the imparting of knowledge, these animals are treated as mere means to certain ends—ends chosen by individual scientists, usually with society's blessing. The animals are spoken of as "tools," "models," and the like, a vocabulary that even to this day continues to reflect the Cartesian view that nonhuman animals are "nature's machines," devoid of all thought and feeling. Some efforts are sometimes made, to be sure, to avoid and alleviate animal pain, but there is no law (at least not in the United States) that requires the use of anesthetic once an animal is in a scientific context; the decision to use or not to use anesthetic legally is left to the discretion of the investigator(s). And, of course, there is no law that protects nonhuman animals in laboratories against an untimely death.

The proof of this (were proof to be required) is that although hundreds of millions of animals are killed in the name of science every year, no scientist has ever been convicted of violating any state or federal statute for killing an animal in the name of science.

The practical problem this customary practice in science poses for ethic-of-care feminism, I think, is that this practice seems to run directly counter to what an ethic of care, supplemented by appeals to logic, would sanction. For if humans are not "distinct from the animal and superior to it," then ethic-of-care feminism, when bolstered by consistency arguments, must demand that we take our moral equality with our animal kin seriously. And this in turn demands, I believe, that we not allow to be done to these animals "in the name of science" what we would not allow to be done to one of our own—in particular, one of those humans with whom we already have a caring, interpersonal relationship. Thus, since we most certainly would not permit a scientist to harm one of our friends, or a spouse, or a parent, or one of our children, treating them as a "tool" or "model" in the hope of benefiting others, I think we must conclude that science, if it were to be guided by the logically enriched moral principles of ethic-of-care feminism, would not permit this way of interacting with nonhuman animals either. In this sense, and for these reasons, ethic-of-care feminism must stand in favor of the total abolition of the harmful use of animals in science.

Whether this finding is true or not, it certainly seems provocative. Comparatively few scientists will welcome it (though the number of scientific antivivisectionists grows daily). And perhaps few feminists will welcome it (liberal feminists, for example, might claim that we "lower" the importance of being human if we argue for moral equality between humans and other animals). And yet this is the conclusion I think we should reach if (1) we accept the view that the dismantling of patriarchy requires that we attack it at its roots, if (2) part of these roots includes the view that humans are "distinct from the animal and superior to it," and if (3) a feminist ethic can allow a place for reason in ethics, especially in the form of consistency arguments, and thus can overcome the moral conservatism that seems to be indigenous to an ethic of care. All these points—(1) through (3)—are contentious, and none admits of simple proof or even simple reasonable defense. But having admitted, indeed, having insisted on the problems we face in these troubled waters, I think we are right to conclude that just as nature is said to abhor a vacuum, so must an ethic-of-care feminism abhor vivisection.

"Politicizing" Science

Defenders of the status quo in science are unlikely to look with favor at the prospect of having ethic-of-care feminists in high places. One can even anticipate the charge that to force science to conform to the ethic of care, as this applies to nonhuman animals, would be to "politicize" this "apolitical" inquiry into truth. And this in turn engenders the specter of "outside interference" in the work of scientists.

Now, I think we know enough to recognize propaganda when we see it, so we will not be easily taken in by appeals to the "value-neutrality" of science or scientists. But the charge that we run the risk of "politicizing" science, and the fear of "outside interference," deserve at least brief comment.

The plain fact is, science *already* is politicized. As the biologist Lynda Birke notes, "scientists are not at present free to pursue any line of inquiry in any way that they wish: public opinion and, to a much greater extent, sources of available funding, impose constraints upon what can be done. A woman wishing to do research relevant to feminist interests, for example, is quite likely to find it difficult to obtain funding, institutional support, laboratory space, or whatever. Yet a scientist doing military research of dubious benefit to humankind, is quite likely to obtain ample funding and facilities from defense budgets."[8]

What is true in the case of researching feminist topics is no less true in the case of developing alternatives to the use of animals in science. Consider the available support for developing such alternatives in the area of education. As the U.S. Congress's Office of Technology Report, *Alternatives to Animal Use in Research, Testing and Education,* states, "in the long run, the most serious problem may be the lack of professional academic rewards for faculty members working in this area. Promotion, tenure, and salary increments are awarded predominantly for productivity in the research laboratory, not for efforts to develop innovative teaching techniques and materials."[9] Is science "apolitical"? Let us have an end to this breezy propaganda. The worry (some) scientists have is not that science, were it to be guided by feminist values, would become "politicized" (and thus "interfered with"), but that the "politics" would be the wrong ones. And that, of course, is an entirely different story, one that requires a careful assessment of the arguments at hand. Whatever their merit, we can at least agree with Lynda Birke when she writes that "the cry of 'interference!' raised when the issue of public accountability [of science] is brought up is of course nonsense, since science already is political, and in that sense, science already is 'interfered' with."[10]

Conclusion

I have argued that ethic-of-care feminism supports the abolitionist position in the case of the harmful use of animals in science. In claiming this I reach a conclusion some feminists have reached—Frances Power Cobbe, for example, who was a major figure in both the suffragette and antivivisection movements in Victorian England, and Coral Lansbury, Carol Adams, and Roberta Kalechofsky among our contemporaries. Still, comparatively few of today's leading feminist thinkers and activists have joined Cobbe and the others in this conviction, which is all the more surprising given the initial role women played in the antivivisection movement, on the one hand, and, on the other hand, given the disproportionately large number of women, in comparison with the number of men, who are part of the estimated ten million people comprising the animal-rights movement, broadly conceived.

Vivisection's nineteenth-century apologists were only too happy to heap scorn on Frances Power Cobbe and other women because of the excessive "sentimentality" that, they charged, characterized "this swarm of buzzing idlers," these "silly women led astray."[11] What could explain why so many women, and so few men, would "bray persistent" against those noble scientists who labored on in their quest for nature's "hidden gifts of cure"? The French physiologist Elie de Cyon has an answer. "Is it necessary to repeat that women—or rather, old maids—form the most numerous contingent of this group? Let my adversaries contradict me, if they can show me among the leaders of the agitation one girl, rich, beautiful, and beloved, or some young wife, who has found in her home the full satisfaction of her affections."[12] In caring for animals, in other words, women express their frustration at not being cared for enough themselves. By men, of course.

Elie de Cyon at least has the virtue of wearing his patriarchal prejudices on his sleeve, which is more than can be said of many. But the same kinds of prejudice toward women we find in Elie de Cyon are to be found today in the prevailing attitudes in science toward nonhuman animals. Or so I believe and have argued. The great pity is too few feminist thinkers and activists have seen the connection, too few are adding their voice to that of Alice Walker, who recognizes in the oppression of nonhuman animals in the lab the same pattern of oppression women face all over the world. Too few have seen this, and done this, up to now, that is.[13]

6 *Abolishing Animal Agriculture*

Time was when philosophers had little good to say about animals other than human beings. "Lawless beasts," writes Plato. "Of the nature of sticks and stones," opines the nineteenth-century Jesuit W. D. Ritchie. True, there have been notable exceptions at least as far back as Pythagoras, who advocated vegetarianism on ethical grounds. Cicero, Epicurus, Herodotus, Horace, Ovid, Plutarch, Seneca, Virgil: hardly a group of "animal crazies," but ethical vegetarians all. By and large, however, a dismissive sentence or two sufficed or, when one's corpus took on grave proportions, a few paragraphs or pages. Thus we find Kant, for example, by all accounts one of the most influential philosophers in the history of ideas, devoting almost two full pages to the question of our duties to nonhuman animals, while St. Thomas Aquinas, easily the most important philosopher–theologian in the Catholic tradition, bequeaths perhaps ten pages to this topic.

Times change. Today an even modest bibliography of the past decade's work by philosophers devoted to the moral status of animals would easily equal the length of Kant's and Aquinas's treatments combined, a quantitative symbol of the changes that have taken place, and continue to take place, in philosophy's attempt to excise the cancerous prejudices lodged in the anthropocentric belly of Western thought.

With relatively few speaking to the contrary (St. Francis of Assisi always comes to mind in this context) theists and humanists, rowdy bedfellows in most quarters, have gotten along amicably when discussing questions about the moral center of the terrestrial universe. *Human* interests form the center of this universe. Let the theist look hopefully beyond the harsh edge of bodily death, let the humanist denounce, in Freud's terms, this "infantile view of the world," at least the two could agree that the moral universe revolves around us humans: our desires, our needs, our goals, our preferences, our love for one another. The intense dialectic now characterizing philosophy's assaults on the traditions of humanism and theism, assaults aimed not only at the traditional account of the moral status of nonhuman animals but at the foundations of our moral dealings with the natural environment, with nature generally—these assaults should not be viewed as local skirmishes between obscure academicians each bent on occupying a deserted fortress. At issue are the validity of alternative visions of the scheme of things and our place in it. The growing philosophical debate over our treatment of the planet and the other animals with whom we share it is both a symptom and a cause of a culture's attempt to come to critical terms with its past as it attempts to shape its future.

I shall limit myself to three of the challenges being raised against moral anthropocentrism. The first comes from *utilitarians;* the second from proponents of *animal rights;* and the third from those who advocate a *holistic ethic.* This chapter offers brief summaries of each position with special reference to how they answer two questions: (1) Is vegetarianism required on ethical grounds? (2) Judged ethically, what should we say, and what should we do about, commercial animal agriculture? To ask whether vegetarianism is required on ethical grounds is to ask whether there are reasons other than those of self-interest (for example, other than those that relate to one's own health or financial well-being) that call for leading a vegetarian way of life. As for the expression "commercial animal agriculture," this refers to the practice of raising animals to be sold for food. The ethics of other practices that routinely kill other animals (for example, hunting and trapping) will not be considered except in passing, not because they are immune to moral criticism but because space and time preclude my offering it here. Space and time also preclude my offering sustained critical assessments of the views I discuss.

Moral Anthropocentrism

Aquinas and Kant speak for the anthropocentric tradition. This tradition does not issue a blank check when it comes to how humans may treat other animals. Positively, we are enjoined to be kind to them; negatively, we are prohibited from being cruel. But we are not enjoined to be the one and prohibited from being the other because we owe such treatment to these animals *themselves.* For we have no duties *to other animals,* according to the anthropocentric tradition; rather, it is because of ***human*** interests that we have those duties we do. "So far as animals are concerned," writes Kant, "we have no direct duties. . . . Our duties to animals are merely indirect duties to mankind." "He who is cruel to animals becomes hard also in his dealings with men," writes Kant. *That* is why cruelty to animals is wrong. As for kindness, Kant observes that "tender feelings towards dumb animals develop humane feelings towards mankind." And *that* is why we have a duty to be kind to animals.[1]

So reasons Kant. Aquinas predictably adds theistic considerations but the main story line is the same, as witness the following representative passage from his *Summa Contra Gentiles.*

> Hereby is refuted the error of those who said it is sinful for a man to kill dumb animals: for by divine providence they are intended for man's use in the natural order. Hence it is not wrong for man to make use of them, either by killing, or in any other way whatever. . . . And if any passages of Holy Writ seem to forbid us to be cruel to dumb animals, for instance to kill a bird with its young: this is either to remove men's thoughts from being cruel to other men, and lest through being cruel to animals one becomes cruel to human beings: or because injury to an animal leads to the temporal hurt of man, either of the doer of the deed, or of another: or on account of some [religious] signification: thus the Apostle expounds the prohibition against *muzzling the ox that treadeth the corn.*[2]

To borrow a phrase from the twentieth-century British philosopher Sir W. D. Ross, our treatment of animals for both Kant and Aquinas is "a practice ground for moral virtue." The *moral game* is played between human players or, on the theistic view, human players plus God. The way we treat other animals is a sort of moral warm-up—character calisthentics, as it were—for the real game in which these animals themselves play no part.

Utilitarianism

The first fairly recent spark of revolt against moral anthropocentrism comes, as do other recent protests against institutionalized prejudice, from the pens of the nineteenth-century utilitarians Jeremy Bentham and John Stuart Mill. In an oft-quoted passage Bentham enfranchises sentient animals in the utilitarian moral community by declaring, "The question is not, Can they *talk?*, or Can they *reason?*, but, Can they *suffer?*"[3] And Mill goes even further, writing that utilitarians "are perfectly willing to stake the whole question on this one issue. Granted that any practice causes more pain to animals than it gives pleasure to man: is that practice moral or immoral? And if, exactly in proportion as human beings raise their heads out of the slough of selfishness, they do not with one voice answer 'immoral' let the morality of the principle of utility be forever condemned."[4]

Some of our duties are *direct duties to other animals,* not indirect duties to humanity. For utilitarians, these animals are themselves involved in the moral game.

Viewed against this historical backdrop the position of the influential contemporary moral philosopher Peter Singer can be seen to be an extension of the utilitarian critique of moral anthropocentrism. In Singer's hands utilitarianism requires that we consider the interests of everyone affected by what we do, and also that we weigh equal interests equally.[5] We must not refuse to consider the interests of some people because they are Catholic, or female, or black, for example. *Everyone's* interests must be considered. And we must not discount the importance of equal interests because of whose interests they are. Everyone's interests must be weighed *equitably.* Now, to ignore or discount the importance of a woman's interests *because she is a woman* is an obvious example of the moral prejudice we call sexism, just as to ignore or discount the importance of the interests of blacks (or native Americans, Chicanos, etc.) are obvious forms of racism. It remained for Singer to argue, which he does with great vigor, passion, and skill, that a similar moral prejudice lies at the heart of moral anthropocentrism, a prejudice that Singer, borrowing a term coined by the English author and animal activist Richard Ryder, denominates *speciesism.*[6]

Like Bentham and Mill before him, therefore, Singer denies that humans are obliged to treat other animals equitably in the name of the betterment of humanity and also denies that acting dutifully toward these animals is a warm-up for the real moral game played between humans or, as theists would add, between humans-and-humans-and-God. *We owe it to those animals who have interests to take their interests into account, just as we also owe it to them to count their interests equitably.* In these respects we have *direct* duties to them, not indirect duties to humanity. To think otherwise is to give sorry testimony to the very prejudice—speciesism—Singer is intent upon silencing.

Animal Agriculture Today

Singer believes that one of speciesism's most obvious symptoms is that we eat other animals, and his utilitarian case for vegetarianism gains strength from the radical changes that recently have taken place in commercial animal agriculture. Increasingly animals raised for food never see or smell the earth. Instead they are raised permanently indoors in unnatural, crowded conditions—raised "intensively," to use the jargon of the animal industry—in structures that look for all the world like factories.

Indeed, it is now common to refer to such commercial ventures as *factory farms.* The inhabitants of these "farms" are closely confined in cages, or stalls, or pens, living out their abbreviated lives in a technologically created and sustained environment: automated feeding, automated watering, automated light cycles, automated waste removal, automated whatnot. And the crowding: as many as nine hens in cages that measure 18 by 24 inches, veal calves confined to 22-inch-wide stalls, pregnant hogs confined in tiers of cages, sometimes two, three, or four tiers high. Many of the animals' most basic interests are ignored, and most are undervalued.

Add to this sorry tale of speciesism on today's factory farms the enormous waste that characterizes the animal industry, waste to the tune of six or seven pounds of vegetable protein to produce a single pound of animal protein in the case of beef cattle, for example; and add to the accumulated waste of nutritious food the chronic need for just such food throughout the countries of the Third World, whose populations characteristically are malnourished at best and literally starving to death at worst; add all these factors together and we have the basis on which utilitarians can answer our two questions. In response to the first question, Is vegetarianism required on ethical grounds? utilitarians can reply that it is, noting that it is not for self-interested reasons alone that we should stop eating the flesh of dead animals (what we call "meat") but for reasons that count the interests of *other* humans and *other* animals. And as for our second question, the one that asks what we should think and do about commercial animal agriculture as it exists today, utilitarians can use these same considerations to support their moral condemnation.

Some Problems for Utilitarianism

Some utilitarians, then, offer an important critique of commercial animal agriculture and, allied with this, an impassioned defense of ethical vegetarianism. Moreover, they do this by appealing to ideas—a principle of equality, and the ideal of making life better for both humans and other animals—that have the ring of profound truth about them. It cannot be right to treat others inequitably or to make their life miserable when this can be avoided. If we know anything, we know this. We owe a lasting debt to all utilitarians for reminding us of these important truths.

As powerful and plausible as utilitarianism is, however, it does not seem to be an altogether satisfactory moral position. Because it is a future-oriented ethic, one that requires that we act now to bring about the best results in the future, it seems to undervalue obligations we have acquired in the past. The obligation to keep a promise is a case in point. We are relieved of this obligation, according to utilitarians, if the results of breaking our word to one person will cause better consequences for others. To follow this direction would make a promise not worth the words with which it was spoken. It should not be easy to go back on one's word. Any theory that implies that it is, is to that degree unreliable. Because utilitarianism does have this implication, our confidence in its adequacy should be shaken.

This is not the only difficulty utilitarians must face. Consider flat-out lying, or even murder. If others will gain more than you would lose if I lie to you, or if I deliberately end your life, then utilitarianism implies that I would do no wrong in either case. And this seems as plainly false as any moral claim can be.

Utilitarianism also is open to the objection that it could sanction the worst kinds of oppression, including a system of indentured slavery. Of course, the theory will not allow us to ignore the interests of slaves, or to count their interests for less than the equal interests of their owners. Having abided by these rules, however, we still need to ask about the consequences: *Would it be better* to permit enslaving a few so that the many might prosper? This question remains entirely open, given the utilitarian position. Indeed, were a plausible case made for finding the overall consequences better, utilitarianism actually would justify this otiose practice.

Now it is true that we often do not know much about consequences, whether those that flow from our individual acts or those stemming from social practices. Human ignorance is more prodigious than human knowledge. This is a plain, and often a regrettable, fact. But it is a fact that gives no comfort to utilitarians. Just the opposite: It adds to rather than subtracts from their problems. Since we know very little about the future, utilitarianism seems to imply that we know very little about what is right and wrong, an implication that blunts the utilitarian attempt to denounce individual wrongs like murder and social evils like slavery. One does not explain why these things are wrong by saying that, because we know so little about the future, they may not be right.

Utilitarians have responses to these and related objections. Most moral philosophers find some or all of them unconvincing. Most moral philosophers, that is, are not utilitarians. But whatever the merits of utilitarianism, considered in general, its particular critique of anthropocentrism deserves our lasting applause. Some of our duties are ***direct duties to other animals.*** Utilitarians have shown that the anthropocentric emperor has no clothes. This much is clear. And it is also clear that commercial animal agriculture must be wrong if it systematically fails to consider the interests of nonhuman animals or fails to count their interests equitably. If this much can be shown—and Singer prosecutes this case persuasively—then the utilitarian case for vegetarianism, as well as its case against commercial animal agriculture, are strong indeed, even if the theory in general is weak.

The Rights View

An alternative to the utilitarian attack on anthropocentrism, and one that also issues its own severe critique of commercial animal agriculture, is the rights view. Those who accept this view hold that (1) certain individuals have moral rights, and (2) these individuals have these rights independently of considerations about the value of the consequences. The rights view, then, is strongly opposed to utilitarianism. As such, it is not open to the most telling objections raised against this latter view. It matters not, for example, how much the many benefit from oppressing the few. Such oppression is and must be wrong, according to the rights view, because it systematically violates the basic rights of the oppressed.

For reasons given elsewhere in this volume I believe that many nonhuman animals are subjects of a life, have inherent value, and thus share with us an equal right to be treated with respect.[7] Morally important questions about line drawing must be addressed in this quarter just as in other morally troubling contexts. But wherever one draws the line with reason, the animals raised for human consumption are on "our" side of it. They are, that is, individuals with a biography, not merely a biology. Those who advocate the rights view, therefore, believe that these animals possess the fundamental moral right to be treated with respect.

Animal Rights and Ethical Vegetarianism

To view these animals in this way makes a truly profound difference to our understanding of what we may do to them. Because biographical individuals have a moral right to respectful treatment, we must never reduce their moral status to that of being useful means to our ends. That being so, the rights view calls for the total dissolution of commercial animal agriculture. Not merely modern intensive rearing methods must cease. For though the harm visited upon animals raised in these circumstances is morally wrong, its removal would not eliminate the more basic wrong its presence compounds. The *basic* wrong is that animals raised to be eaten are treated disrespectfully, as if they were "commodities," "economic units," "investments," "a renewable resource," and so on. In fact, they are biographical individuals and so, like us, are owed treatment that accords with the right to be treated with respect, a respect we fail to show when we end their life so that we might eat them.

Some people think the rights view is too "radical" or "extreme," calling as it does for the total abolition of a culturally accepted institution. But when an institution is grounded in injustice, as animal agriculture, according to the rights view, is, then there is no room for internal house cleaning. Morality will not be satisfied with anything less than its total abolition. And that, for the reasons given, is the rights view's verdict regarding commercial animal agriculture.

Some Problems for the Rights View

Not the least of the rights view's strengths already has been noted: namely, its immunity from the most telling objections urged against utilitarianism. The end does not justify the means. The rights view categorically prohibits violating the rights of some so that others might benefit. No biographical individual—not for reason of race, or sex, or species—exists as somebody else's resource. We are not to do evil that good might come, not to violate the rights of the individual in order to help others. Whatever gains such injustice yields are ill-gotten. Morally, then, they should not be gotten at all.

Like utilitarianism, however, the rights view is not free of problems. One concerns possible conflicts of rights, cases, for example, where in the nature of the case we cannot avoid harming someone. Whatever we do, *some* evil will result. What ought we to do in such cases? And how can we decide this without having recourse to utilitarian considerations? It is not easy for advocates of the rights view to answer these questions, though answer them they must. If the moral life were void of such conflicts, moral theory would be vastly simpler. But the moral life is not free of such conflicts, which in part is why moral theory is so difficult. One thing is clear: We do not solve these difficulties just by appealing to "the rights of the individual," as if this venerable idea were a magic wand.[8]

A different challenge probes the meaning of respect. For the most part advocates of the rights view interpret this idea negatively. If we are to show respect for the individual, we must not treat the individual as a resource, must not treat the individual as a commodity, and so on. Is there a positive side to this idea? Do we do everything we should merely by doing nothing? Or must we also actively help those who are harmed by others?

To some degree the rights view already has answered this question. When individuals are the victims of *human injustice,* we have a duty to help the victims. This is well and good, as far as it goes. But is that all there is? Do we owe nothing more? What of cases where the victims are not harmed because of human injustice but because of natural catastrophes—earthquakes, floods, and tornadoes, for example? Do we have a duty to help these human beings? If we do, is it possible for the rights view to explain this duty without having recourse to (what for them are forbidden) utilitarian considerations?

A more difficult question changes the plot only slightly. Suppose *some nonhuman animals* are being harmed by *other nonhuman animals* (for example, in predator–prey relationships). Do we have a duty to help? If so, which ones? The predators? Or the prey? Clearly, even if it is true, as I believe it is, that the rights view has answered some basic moral questions both correctly and decisively, there are many other important questions its advocates must address, and address fairly. Otherwise this philosophy, like any other, runs the risk of reducing itself to flag-waving propaganda.

Holism

The abolitionist implications of both the utilitarian and the rights view's critique of commercial animal agriculture suggest how far some philosophers have moved from the anthropocentric traditions of theism and humanism. Like utilitarian attacks on this tradition, however, the rights view seeks to make its case by working within the major ethical categories of this tradition. For example, utilitarians do not deny the moral relevance of human pleasure and pain, so important to our humanist forbearers; rather, they accept this and seek to extend our moral horizons to include the moral relevance of the pleasures and pains of other animals. And the rights view does not deny the moral importance of the individual, a central article of belief in theistic thought; rather, it accepts this moral datum and seeks to widen the class of individuals who are thought of in this way to include many nonhuman animals.

Because both the positions discussed in the preceding use major ethical categories handed down by our predecessors, some influential thinkers argue that these positions, despite all appearances to the contrary, remain in bondage to anthropocentric prejudices. What is needed, these thinkers believe, is not a broader interpretation of traditional categories (for example, the category of "the rights of the individual"), but the overthrow of these very categories themselves. Only then will we have a new vision, one that liberates us from the last vestiges of anthropocentrism.

Among those whose thought moves in this direction none is more influential than Aldo Leopold's.[9] Leopold rejects the individualism so dear to the hearts of those who build their moral thinking on "the value (or rights) of the individual." What has ultimate value is not the individual but the collective, not the part but the whole, meaning the entire biosphere and its constituent ecosystems. Acts are right, Leopold believes, if they promote the integrity, beauty, diversity, and harmony of the biotic community; they are wrong if they work against these values. As for individuals, be they humans or animals, they are merely "members of the biotic team," having neither more nor less value in themselves than any other member—having, that is, no value in themselves. What value individuals have, so far as this is computable at all, is instrumental only: They are good to the extent that they promote the values of the biotic community.

Utilitarianism, not just the rights view, goes by the board given Leopold's vision. To extend our moral concern to the pleasures and pains of other animals is not to overcome the prejudices indigenous to anthropocentrism. One who does this is still shackled to these prejudices, supposing that those mental states that matter to humans must be the measure of what matters morally to the world at large. Utilitarians are people who escape from one prejudice (speciesism) only to embrace another (what we might call "sentientism"). Animal liberation is not nature liberation. In order to forge an ethic that liberates us from our anthropocentric tradition we must develop a holistic understanding of the community of life and our place in it. "The land" must be viewed as meriting our equal moral concern. Waters, soils, plants, rocks—inanimate not just animate existence—must be seen to be morally considerable. All are equal members of the same biotic team.

Holism and Individual Animals

Holists face daunting challenges when it comes to determining what is right and wrong. These are to be determined by calculating the effects of our actions on the life community. Such calculations will not always be easy. Utilitarians, as was noted earlier, encounter a serious problem when they are asked to say what the consequences will be if we act in one way rather than another. And this problem arises for them despite the fact that they restrict their calculations to sentient life. How much more difficult it must be, then, to calculate the consequences for the entire *biosphere!*

But perhaps the situation for holists is not as dire as I have suggested. While it is true that we often lack detailed knowledge about how the biosphere is affected by human acts and practices, we sometimes know enough to say that some of the things we are doing are unhealthy for the larger community of life. For example, we do not know exactly how much we are contaminating the water of the earth by using rivers and oceans as garbage dumps for toxic wastes, or exactly how much protection afforded by the ozone layer is being compromised by our profligate use of chlorofluorocarbons. But we do know enough to realize that neither situation bodes well for marine and other life forms as we know them. Those scientists and policymakers who insist on having "more data" before they raise their hands to vote in favor of serious social change are like the passersby who could have prevented a flood if they would have but lifted a finger.

Let us assume, then, what I believe is true, that we sometimes are wise enough to understand that the effects of some human practices act like insatiable cancers eating away at the life community. From the perspective of holism, these practices are wrong, and they are wrong because of their detrimental effects on the interrelated systems of biological life.

It is important to realize that holists are aware of the catastrophic consequences toxic dumping and the ever-widening hole in the ozone layer are having on individual biographical beings—on seals and dolphins, for example. And it would be unfair to picture those who subscribe to holism as taking delight in the suffering and death of these individual animals. Holists are not sadists. What is fair and important to note, however, is that the suffering and death of these animals are not morally significant according to these thinkers. Morally, what matters is how the diversity, sustainability, and harmony of the larger community of life are affected, not what happens to individuals.

To make the holists' position clearer, consider the practice of trapping fur-bearing animals for commercial profit. Holists find nothing wrong with this economic venture so long as it does not disrupt the integrity, diversity, and sustainability of the ecosystem. Trappers cause such disruptions when they overtrap a particular species. The danger here is that the depletion of a particular species will have a ripple effect on the community as a whole and that the community will lose its diversity, sustainability, and integrity. The overtrapping (and hunting) of wolves and other predatory animals in the northeastern United States often is cited as a case in point, though not very convincingly in my view.[10] Once these natural predators were removed, other species of wildlife—deer in particular—are said to have over-populated, so that today these animals actually imperil the very ecosystem that supports them. All this could have been avoided if, instead of rendering local populations of natural predators extinct by overtrapping and overhunting, they had been trapped or hunted more judiciously, with an eye to sustainable yield. Although a significant number of individual animals would have been killed, the integrity, harmony, and sustainability of the ecosystem would have been preserved. When and if commercial trappers achieve these results, holists believe they do nothing wrong. From the perspective of holism, the inevitable suffering and untimely death of individual fur-bearing animals do not matter morally.

Holism and Ethical Vegetarianism

Holism's position regarding the ethics of vegetarianism is analogous to its position regarding the ethics of commercial trapping. There is nothing wrong with raising animals for food if doing so is good for the larger life community. But it is wrong to do this if the community suffers. For example, raising cattle on nonarable lands arguably adds to the diversity of the ecosystem, and as long as the lands are not overgrazed, the system remains sustainable. However, to destroy delicately balanced communities of life in order to create new grazing lands for commercially raised beef cattle, as is being done in the Amazon rain forest, is wrong. For holists, then, the ethics of meat eating must be judged on a case-by-case basis. What matters most is where the meat comes from, not the pain and death of the animals involved. If these animals are raised in an ecologically sensitive way, we do nothing wrong when we eat them.

While this may seem like good news to meat eaters, it isn't. Speaking generally, commercial animal agriculture is an ecological disaster. Or, rather, it is part of a more general ecological disaster, one that begins with grain production. Almost all the grain grown in affluent countries (over 90 percent of the oats, barley, and rye, for example) is used as animal feed. The agriculture that produces these massive amounts of animal feed is literally killing the planet. Increasingly, knowledgeable people are reaching the conclusion that it is not sustainable. Because the same crop is grown on the same land year after year after year (this is what is meant by saying the system is "monocultural"), and because of the heavy use of toxic synthetic chemicals such as herbicides, nematocides, and fungicides (this is what is meant by saying the system is "chemically intensive"), the future fertility of the land is being compromised.

And it is not just the land that is a casualty of modern agriculture. The quality of our water also is at risk. Once applied to crops, chemicals do not disappear. Often they run off into neighboring creeks or rivers, or trickle down through the earth into underground lakes only to surface again, at some other time and place. We do not speak loosely when we say that chemically intensive, monocultural agriculture is far and away the greatest single cause of the deteriorating quality of the earth's water supply. Because commercial animal agriculture is the largest consumer of a system of grain production that has these deleterious ecological consequences, holists can, and holists are, speaking out in favor of a vegetarian way of life.[11]

There is a longer story to be told about the ecological carnage attributable to commercial animal farming, a *much* longer story than can be told here, one that would detail, for example, the environmental degradation that is the *direct* result of factory farms.[12] Suffice it to say on this occasion that the more we learn about why and how commercial animal agriculture is implicated in environmental degradation, the stronger holism's case in favor of vegetarianism becomes, and this, paradoxically, despite the fact that what is done to the animals meat eaters eat is not morally relevant.

Some Problems for Holism

There is much to be said in favor of holism, and also much to be said against it. By challenging us to consider how our individual and collective behavior impacts on the larger life community, it forces us to question not only the merits of anthropocentrism but also those of utilitarianism and animal rights. Moreover, by tying the standard of ethical conduct to what benefits or harms the planet, holism sees human beings as embedded in the natural world, not as transcendent over it, a view well supported by our best science, including evolutionary biology and ecology in particular. For these reasons alone, holism makes an important contribution to the contemporary debate over what humans are and what we ought to be.

But holism arguably makes an even more important contribution. The principle that ecologically insensitive human practices are wrong has a high degree of plausibility. It also has enormous potential in the battle to preserve and restore the integrity of the planet. What we humans are doing to the forests and the oceans, the air and the land is morally tragic, and our understanding of *why* it is, is at least partially attributable to Leopold and other holists.

But holism goes too far when it maintains that any ecologically sensitive human practice is right. This must be false, if any moral claim is. Judged on the basis of all the available evidence, cotton production in the antebellum South was ecologically sensitive. The ecology of the area was in harmony, the agriculture was sustainable. Yet surely we are not prepared to say that the forced labor of indentured slaves was therefore quite all right. There is more to morality than asking how the life community is affected by what we do. There are also questions that concern the justice of how we do them. Because holists like Leopold look only to ecological ends and neglect the ethics of the means used to achieve them, their position is a half-truth at best, a serious threat to justice at worst.

What is true of holism's inability to criticize ecologically sensitive but unjust practices involving human beings, is no less true of similar practices involving other animals. This can be illustrated by first considering commercial trapping. It is not enough to ask whether commercial trapping is ecologically sensitive, since, like cotton produced by slaves, it might be this and *still* be wrong. We also need to ask (minimally) how this end is achieved.

Among the things we know is this. Some traps used by trappers, such as the infamous steel-jawed leghold trap, are veritable instruments of torture. Most civilized countries have outlawed the steel-jawed leghold trap because of its barbarous cruelty. Holists must find it difficult to add their voice. What matters to them is not the means, and not whether fur-bearing animals suffer, but whether trapping is sustainable. Surely this cannot be right. Quite aside from questions of sustainable populations, the suffering of individual trapped animals matters morally, just as, quite aside from questions about sustainable agriculture, the suffering of human slaves matters morally, too.

An analogous argument can be given in the case of commercial animal agriculture. Even when this form of agriculture is ecologically sensitive (increasingly rare in this day of factory farms), it does not follow that it is right. We must also ask, How is this end achieved? What are the means? The means, in the simplest terms, are: by bringing about the untimely death of biographical animals. Does this matter morally? Not to holists. But, then, neither does the untimely death of human beings who, by their forced labors, contribute, like the good "team members" they are, to ecological sustainability. On a previous occasion I characterized the sort of holism we find in Leopold as "environmental fascism."[13] This indictment remains. Like political fascism, where the individual is made to serve the interests of the larger political community, an unbridled ecological holism, where it is permissible to force the individual to serve the interests of the larger life community, is fascistic too.

Defenders of holism might be tempted to say that there are morally relevant differences between human oppression and the oppression of other animals. These defenders more than have their work cut out for them, first, because holism does not offer adequate protection even to human beings (recall holism's inability to criticize a system of indentured slavery when it is ecologically sensitive), and, second, because the clamoring voices of utilitarians and partisans of the rights view, with their cries of "Speciesism!" will not be easily silenced. Whether holists can answer these thinkers adequately is too large a question to be settled here. It is enough to understand why holism's critics remain unconvinced that ecological sensitivity is a satisfactory standard of what is right in general, and of our treatment of other animals in particular.

Sameness Among the Differences

Philosophers want to be right, and they want to be right for the right reasons. In the nature of the case, therefore, philosophy is an argumentative discipline. The way one philosopher shows respect for another philosopher is by *challenging* the other's ideas, by trying to show (fairly, of course) that these ideas are false, or unsupported, or worse. To do anything else would be . . . well, philosophical bad manners. And so it is that, in this chapter as in others, I have pointed out not only what seem to me to be the strengths but also what seem to me to be the weaknesses of various positions.

In the broadest terms my conclusions in this essay are as follows. Sometimes I think the views I have discussed are right, sometimes for what I think are the right reasons. But not always. Sometimes I think they are wrong on both counts, and sometimes I think their mistakes are far from trivial. We should not take lightly views that imply, for example, that slavery just might possibly be right. We should not take this lightly at all.

Still, it is important not to miss seeing the forest for the trees, not to overlook important similarities because we become fixated on important differences. (Philosophers, myself included, have a remarkable tendency to do this.) In order that I might make wrong impressions right, I want to close by mentioning one of the important similarities unearthed along the way.

Most people who read this book, and who still eat meat, will buy the meat they eat at nationally franchised food stores. This meat is therefore the end product of factory farms. How we know this is simple. It's a matter of economics. Wholesalers and retailers buy as cheap as they can, which means—in the case of meat—at the trough of factory farms. It's that simple.

Now, despite their many differences, proponents of these three philosophies—utilitarians, holists, and advocates of the rights view—can speak with one voice on some occasions. This is one of them. All agree that factory farming is wrong: wrong because it violates the rights of individual animals (the position of the advocate of the rights view), wrong because it either does not count the interests of nonhuman animals at all or does not count them equitably (the utilitarian's position), or wrong because it is destroying the ecology of the planet (the holist's position). Cut it any way you wish, these influential philosophies agree: This form of animal agriculture is wrong.

If the system of supply is wrong, what about individual consumers? Can these three views, despite their many differences, reach the same conclusion here too? They can. Viewed against the backdrop of factory farming, each can say the same thing (though for different reasons): *Vegetarianism is ethically required.* And that is enough for the rest of us to digest today. More than enough, really, since to unburden one's self of the weight of culturally conditioned prejudices (and a culturally encouraged palate) is not easy. For a while, then, it is enough to act on the conclusion these philosophers have put on our plates: Give up eating meat. After some good vegetarian food we can ask which one of these three philosophies has the right reasons. There is time enough to ask many interesting questions with our words, but limited time to answer some of the more important ones with our deeds.

7 *Irreconcilable Differences*

Until quite recently public concern about "the environmental crisis" had gone out of style. This contrasted sharply with fashions back in the late sixties and early seventies. Back then talk about "the environmental crisis" was all the rage. Earth Day was celebrated annually to remind everyone about how bad things were going for the planet, and President Jimmy Carter even went so far as to declare that the fight to save the environment was "the moral equivalent of war."

That war was never won. In fundamental ways it was never fought. The problems President Carter recognized are still with us: air and water pollution, destruction of natural habitats, erosion of topsoil, extinction of species, for example. But today there is a difference. Today the list of problems is longer: the destruction of the ozone layer, acid rain, the plowing under of the rain forest, changes in the earth's temperature, the gradual encroachment of the oceans, the disposal of toxic wastes. All one need do is read a newspaper or watch the evening news to realize the variety and importance of the environmental challenges we face. If it was true back then, in the late sixties and early seventies, that we faced an environmental crisis, it is more true today. That not enough of us recognize the crisis for what it is or, seeing it, do not care enough to do anything about it, only helps make a bad situation worse.

One other thing has changed in the past twenty years, something that might give us cause for hope. This is the seriousness with which academic moral philosophers take the environmental crisis. Twenty years ago it would have been unusual for an academic philosopher to write or lecture about such real-life problems as pollution or species extinction. Not so today. Now there are even professional journals devoted exclusively to environmental questions, and the number of books published in the area of environmental ethics during the past twenty years easily dwarfs the number published in the previous two thousand. To their credit, philosophers are not fiddling while the Earth burns.

Among the questions these thinkers are considering, the most basic one asks how the environmental crisis could have happened in the first place. In other words, fundamentally considered, what are its underlying causes? As one should expect, a variety of provocative answers has been offered. One holds that the fault should be laid squarely at the door of the Judeo–Christian tradition, because of its declaration of human "dominion" over nature. Another locates the root cause in the free-market approach to our interactions with nature favored by capitalism, an approach that measures all values, including whatever people find valuable in nature, according to some monetary yardstick.

Each of these and other possible answers deserves our best thought. For obvious reasons, I cannot consider all of them on this occasion. Here I must be more selective and limit my attention to only some of the possible answers to the question at hand: Fundamentally considered, what are the root causes of the environmental crisis?

Among those philosophers and other writers who offer an answer to this question is a group of notable feminist thinkers (so-called ecofeminists), and one thing I want to do is describe and critically assess, though of course incompletely, some of the contributions a few of these thinkers are endeavoring to make. I also want to explore how close, and yet how far away, ecofeminists are from another philosophical tendency in environmental ethics, the tendency now commonly referred to as "deep ecology." What I hope to be able to explain is why, despite their important points of agreement, these two tendencies in environmental philosophy must encounter a number of irreconcilable differences. We do not, therefore, if I am right, have the luxury of being able to embrace both these tendencies at the same time. Instead, we must recognize the difficulty of accepting either the one, or the other. Or possibly neither. My own considered preference will become clearer as we proceed.

In order to understand and appreciate these matters I begin by discussing Carolyn Merchant's 1980 book, *The Death of Nature: Women, Ecology, and the Scientific Revolution.* This is a safe, neutral place to begin, I think, since both ecofeminist philosophers and deep ecologists accept many of this work's most important teachings.

The Death of Nature

In Merchant's widely discussed work she paints the broad outlines of the history of science viewed from a feminist perspective. Speaking generally, two opposing views of the world can be distinguished. The first, firmly rooted in the ancient world and in such "primitive" cultures as those we find among native Americans, is organic and holistic—organic in that the cosmos is perceived as and understood to be an organic entity, "growing and developing from within, in an integral unity of structure and function,"[1] and holistic because all that exists is perceived as and understood to be interconnected, like the limbs, organs, and other components of the human body. "Each portion of an ecological community, each niche, exists in a dynamic relationship with the surrounding ecosystem. The organism occupying any particular niche affects and is affected by the entire web of living and nonliving components."[2]

The second major worldview, associated with the scientific revolution of the seventeenth century, is mechanical and atomistic—atomistic because the world is understood to be populated by discrete, irreducible individuals, each ontologically distinct and separate from every other, and mechanistic because the behavior of these inert ontological atoms, no one of which is capable of moving as it wills (for none has a will), is understood to be governed by the application of mathematically precise laws from outside rather than by teleological aspirations from within. In Merchant's view it is this mechanical view that is "now taught in most Western schools [and] accepted without question as our everyday common sense reality."[3]

Both conceptions—the organic and the mechanical—have been presented in the imagery of the female. "Central to the organic theory," Merchant writes, "was the identification of nature, especially the earth, with a nurturing mother: a kindly beneficent female who provided for the needs of mankind in an ordered, planned universe."[4] Evidence for the feminine aspects of the organic conception of nature comes from diverse sources—for example, mining, where minerals and metals were seen as "ripen[ing] in the uterus of the Earth Mother, [and where] mines were compared to her vagina,"[5] and theology, where we find this representative passage from Richard Hooker: "God being the author of Nature, her voice is but his instrument. By her from him we receive whatsoever in sort we earn."[6] For Hooker and his contemporaries the voice of nature was a feminine voice.

That voice did not always speak the language of the nurturing mother. Along with nature-as-female-nurturer the organic view also recognized the unruliness and destructive potential of nature/woman: "wild and uncontrollable nature . . . [the source of] violence, storms, droughts, and general chaos."[7] Nature/woman sometimes needed to be tamed and subdued, and this was, at least in the dominant traditions of the West, a man's job.

Viewed in this light one can, and Merchant does, interpret the real significance of the scientific revolution as the eradication of one of the feminine dimensions of the organic view and as the validation of the other. The idea of nature as nurturing mother is eradicated totally by the triumph of the view that nature is comprised of lifeless, inert, and ontologically discrete atomic individuals. And any scruples standing in the way to dominating nature, unruly female that she is, were removed by viewing "her" in the impersonal terms mechanism favors: nature/woman as lifeless, inert, passive, and—not least of all—dumb.

"The metaphor of the earth as nurturing mother," Merchant writes, "was gradually to vanish as a dominant image as the Scientific Revolution proceded to mechanize and to rationalize the world view. The second image, nature as disorder, called forth an important modern idea, that of power over nature. Two new ideas, those of mechanism and of the domination and mastery of nature, became core concepts of the modern world. An organically oriented mentality in which female principles played an important role was undermined and replaced by a mechanically oriented mentality that either eliminated or used female principles in an exploitative manner. As Western culture became increasingly mechanized in the 1600s, the female earth and virgin spirit were subdued by the machine."[8] Thus has the conquest of mechanistic science signaled the death of nature (the title of Merchant's book, and its main theme).

Bacon's "New Science"

Among those figures who played key roles in the conceptual transformation of our understanding of nature few made more important or longer-lasting contributions than Francis Bacon (1561–1626). "The father of modern science," as he frequently is called, Bacon was the founder of the inductive method in science and a pioneering thinker in what today we might call the philosophy of technology. However, if we view this much revered and honored thinker from (in Merchant's words) "the perspective of nature, women, and the lower orders of society," she believes "a less favorable image [emerges]."[9] It will be worth our while to consider, though very briefly, some of the evidence Merchant brings against Bacon in this regard.

Merchant offers numerous examples of Bacon's use of his favored imagery: the imagery of subjugation. "The new man of science," in Bacon's words, is not to suppose that the "inquisition of nature is in any part interdicted or forbidden." Nature is to be "bound into service," made a "slave," and put "in constraint" by use of the new mechanical arts. "There is," Bacon writes in a not uncharacteristic passage, "much ground for hoping that there are still laid up in the womb of nature many secrets of excellent use."[10] New discoveries in technology and the invention of gunpowder, for example, were celebrated by Bacon as tools that would help their users "think about the secrets still locked in nature's bosom": "They do not, like the old [means of investigation], merely exert a gentle guidance over nature's course; they have the power to conquer and subdue her."[11] In Merchant's view, what she calls "sexual politics" is compressed into the very language Bacon uses, and it is this same sexual bias, she maintains, that has "helped to structure the nature of the empirical method that would produce a new form of knowledge and a new ideology of objectivity seemingly devoid of cultural and political assumptions."[12]

Further Variations on the Main Theme

Merchant's interpretation of the ascendency of a mechanistic over an organic view of nature is only one, albeit a very important, chapter in the larger saga of the ascendency of masculine over feminine experience and values, as seen by feminist thinkers. This "patriarchal ideology," as it sometimes is called, has tended to celebrate the masculine over the feminine in ways that date back at least to Plato, well before the rise of mechanism. Man is rational (which is good), woman is emotional (which is bad). Man is objective (which is good), woman is subjective (which is bad). Man's end is to transcend nature through reason, woman's fate is to remain in bondage to nature through her physical embeddedness (as the bearer of children) in it.

What gives man his supposed superiority? A number of answers are possible but the general drift always is the same.[13] The male members of the species *Homo sapiens* by nature have greater potential to actualize distinctively human capacities.

Men are (or can be) more rational and less emotional, more objective and less subjective, more impartial and less biased, for example, than women. More generally, it is a person's body and a person's emotional bonds to it that are bad (something to be overcome), while the mind and the love of reason are good (something to be developed as far as possible). Thus, if we grant the unspoken but no less real identification of woman and nature Merchant has attempted to document, then the kinds of behavior masculine superiority and power authorize in the case of a man's dealings with nature will be more readily sanctioned in the case of his dealings with a woman. On this view, nature, no less than women, exists to be tamed, subdued, controlled, and exploited by men.

These large themes—the unspoken identification in Western culture of women with nature, on the one hand, and, on the other, the aspiration among men to transcend nature and subdue it—these large themes have been assimilated into the fabric of an emerging version of feminism now commonly referred to as *ecofeminism.* Those feminists who favor this perspective see common themes in the exploitation of women and nature, and yet tend to celebrate rather than to conceal or minimize the greater closeness women have to nature and natural processes (in particular, the closeness experienced by women as natural producers and nurturers of life). These feminists call for greater care of and empathy with the natural world, which is seen as our shared mother.

Ecofeminists are not the only ones among our contemporaries who call for increased empathy with and care for nature. In particular those philosophers and others associated with what today is called "deep ecology" have joined ecofeminists in calling for a reawakening of relatedness-to-nature. As is true of ecofeminists, deep ecologists also repudiate *both* the (masculine) ideal of transcending nature through reason *and* the mechanistic view of the world with which that ideal has been historically allied. Moreover, like ecofeminist critics of mechanism, deep ecologists favor a return to the organic, holistic view of nature described by Merchant. But whereas feminists, as we have seen, lay the blame for the sins of Western culture at the door of patriarchy, deep ecologists, as we will soon see, offer a different diagnosis.

Deep Ecology

Although the expression "deep ecology" is used in a variety of ways, not all of which are equivalent, it appears to have both a core meaning and a message. In the minds of some the Australian philosopher Warwick Fox expresses both when he characterizes the basic intuition of deep ecology as follows: "It is the idea that we can make no firm ontological divide in the field of existence: That there is no bifurcation in reality between the human and the non-human realms . . . to the extent that we perceive boundaries, we fall short of deep ecological consciousness."[14] That Fox should speak of deep ecological *consciousness* is not unimportant. More than a body of written or spoken propositional truths, deep ecology is a way of *experiencing* the world and our place in it. In this sense—but not only in this sense—deep ecologists liken their search for knowledge to those spiritual quests associated with the devotees of the world's religions: Buddhism, Taoism, Christianity, and native American rituals, for example.

This "new consciousness," as it sometimes is called, is possible to initiates only if they overcome their sense of separateness and superiority. In place of the separate self (the atom of consciousness in a mechanistic world) deep ecology finds the self *as it exists in relation to others,* and in the place of human superiority and transcendence over all that is nonhuman the consciousness awakened to deep ecology experiences equality and immanence. One finds one's self in others not through books and lectures but by awakening to the reality of one's essential relationship to all that is. And this process is fueled by the questing (and questioning) mind which, in its struggle to free itself from illusion, finds respite and insight in moments of quiet meditation.

Both the major themes I have just sketched—first, the theme of a new form of self-realization and, second, the theme of an awakening to biocentric equality—are developed by the Norwegian philosopher Arne Naess, who first coined the expression "deep ecology." In their widely read and discussed book, *Deep Ecology: Living as if Nature Mattered,* Bill Devall and George Sessions offer exemplary summaries of these two "ultimate norms" identified by Naess. To consider the norm of self-realization first, they write as follows:

> In keeping with the spiritual traditions of many of the world's religions, the deep ecology norm of self-realization goes beyond the modern Western *self* which is defined as an isolated ego striving primarily for hedonistic gratification or for some narrow sense of individual salvation in this life or the next. This socially programmed sense of the narrow or social self dislocates us. . . . We are thus robbed of beginning the search for our unique spiritual/biological personhood. Spiritual growth, or unfolding, begins when we cease to see ourselves as isolated and narrow egos and begin to identify with other humans from family and friends to, eventually, our species. But the deep ecology sense of self requires a further maturity and growth, an identification which goes beyond humanity to include the nonhuman world . . . [which] is best achieved by the meditative deep questioning process . . . [and leads to] the realization of 'self-in-Self' where "Self" stands for organic wholeness. This process of the full unfolding of the self can also be summarized by the phrase, "No one is saved until we are all saved," where the phrase "one" includes not only me, an individual human, but all humans, whales, grizzly bears, whole rain forest ecosystems, mountains and rivers, the tiniest microbes in the soil, and so on.[15]

Regarding the second "ultimate norm" of deep ecology, namely, the norm of biocentric equality, Devall and Sessions write as follows:

> The intuition of biocentric equality is that all things in the biosphere have an equal right to live and blossom and to reach their own individual forms of unfolding and self-realization within the larger Self-realization. The basic intuition is that all organisms and entities in the ecosphere, as parts of the interrelated whole, are equal in intrinsic worth. . . . Biocentric equality is intimately related to the all-inclusive Self-realization in the sense that if we harm the rest of Nature then we are harming ourselves. There are no boundaries and everything is interrelated. But insofar as we perceive things as individual organisms or entities, the insight draws us to respect all human and nonhuman individuals in their own right as parts of the whole without feeling the need to set up hierarchies of species with humans at the top.[16]

Devall and Sessions readily concede that these "ultimate norms" are not validated "by the methodology of modern science based on its usual mechanistic assumptions and its very narrow definition of data."[17] But that does not dampen their enthusiasm for their truth. "They [that is, these two norms] cannot be fully grasped intellectually," they write, "but are ultimately experiential."[18] If only we would make ourselves sufficiently open to the realization of "self-in-Self," it seems fair to say, Devall and Sessions believe that these two ultimate norms will be revealed to us.

Similarities

The similarities between the tenets of deep ecology and ecofeminism should be apparent. Both views are united in at least the following ways. Both would insist that

- We are *not* discrete "atoms of self" alienated from the rest of the world; instead, we are part of the living whole, "a member of the biotic team" (as Aldo Leopold expresses the point).
- We are *not* "superior" to the rest of nature. It is *not* our task to "transcend" the natural order through the exercise of our reason, but to *reestablish* our kinship with the whole of nature through a reawakening of our immanence in the natural world.
- The path to self-understanding we are to follow is *not* that of "objective" knowledge, but is the one of subjective (experiential) insight—of feeling, intuition, a sense of the interrelatedness of all that exists.

In these ways, and in others, there is a discernible sense of alliance between deep ecologists and ecofeminists. Superficially considered, the union of these two views would appear to be a proverbial marriage "made in the stars." The question we need to ask is, Is this true? I do not believe it is. For beneath the similarities there lurks, I think, an unbridgeable gulf that separates these two ways of understanding our place in the scheme of things.

For consider: What is it that deep ecologists fundamentally oppose? In one word it is *anthropocentrism:* the belief (roughly speaking) that *human interests* are the measure of all that is valuable. Such a narrow conception of the human's place in the great chain of being stands as the greatest impediment to our quest for self-understanding, deep ecologists believe. We must learn to lose ourselves in others, if we are to find ourselves at all, and this we will never do if we haughtily strut through the universe in the false belief that we are "superior" to everything else.

Ecofeminism also has its diagnosis of the major ills of our culture, but it differs in important ways from the one offered by deep ecologists. The underlying cause is not *human*-centeredness (not anthropocentrism), but *male*-centeredness (androcentrism). Or to express the difference differently, the underlying cause is *not* humanism, it is *patriarchy.* To the extent that deep ecologists misidentify this cause, given the ecofeminist diagnosis, to that extent there is more than ample reason to worry about the durability of a union of the two philosophies.

The Case of Hunting

Just how much reason there is to worry in this regard can be highlighted by contrasting how the two philosophies view a widely entrenched practice in the West—namely, recreational (sport) hunting. Devall and Sessions situate this way of relating to nature in the same category of activities as sunbathing, surfing, sailing, and bicycling: done with the "proper attitude," hunting, like these other activities, is "especially useful" in encouraging the spiritual "maturity" of the self.[19] What, then, is the "proper attitude"?

In an illuminating review of the prohunting literature written by deep ecologists (including Aldo Leopold and the poet Gary Snyder, for example) the philosopher Marti Kheel brilliantly examines the recurring theme of hunting as a ritual means of self-awakening and self-realization. "The mind-set that [these writers] reveal," she observes, "is that of a psychological instrumentalism in which the animal is seen, not as a unique, living being but rather as a means to achieve a desired psychological state. The animal is thus reduced to the status of object or symbol."[20]

Moreover, this psychological state is described in hedonic imagery reminiscent of some of Bacon's finest passages. For example, Ortega y Gasset proclaims that blood has "unequaled orgiastic power."[21] Those who would substitute wildlife photography for hunting, in Ortega's view, might just as well substitute Platonic love for the real thing. As Randall Eaton (a strident advocate of sport hunting) observes, "the hunter's feeling for his prey is one of deep passion [and] ecstacy. . . . The hunter loves the animal he kills."[22]

Eaton, Ortega, Leopold, and other apologists of sport hunting are of one mind when it comes to the naturalness of the desire to hunt. It is "a deep and permanent yearning in *the human condition,*" according to Ortega.[23] Leopold puts the same point only slightly differently: "The instinct that finds delight in the sight and pursuit of game *is bred into every fibre of the human race.*"[24] As Kheel notes—and her point cannot be emphasized too strongly—these patrons of deep ecology are here "describ[ing] hunting not as a necessary means of subsistence, but rather as a *desire* that fulfills a deep psychological need."[25] It is *instinctive* to the species *Homo sapiens* (part of the "human condition," an undeniable strand in "every fibre of the human race") to want to hunt and kill.

On its face, of course, this chorus of celebration of nonsubsistence hunting seems to flatly contradict the second of the two basic norms of deep ecology—so-called biocentric equality. How can it be good and proper to destroy, in the name of sport or recreation, what has the same value as one's self? This is not a simple question for deep ecologists by any means, but the favored response turns on making a distinction between (1) the value of *species as a whole* and (2) the value of *individual members.* If the view one adopts is that it is species as a whole that have value (that are "biocentrically equal" one to the other), then the destruction of species, but *not* the destruction of individuals, emerges as evil. Thus, if one is a species-egalitarian (if, that is, one views all *species* as having the same value), then sport, or recreational, hunting is not destructive of anything valuable, *so long as one does not overhunt*—that is, so long as one leaves a "sustainable yield," as "wildlife managers" are fond of saying. And thus it is, moreover, that what the experience of hunting opens to the hunter is the sense of union with the larger *community* of life, *not* empathy or caring for *individual* animals.

No feminist (and, so, no ecofeminist) will find this vision entirely acceptable. For the gospel of deep ecology is the gospel of patriarchy, this time attired in mystical drag. The familiar themes of power and domination remain intact, only it is wild animals, not wild women, who exist to be vanquished in the name of self-realization. And the familiar theme of transcendence also remains alive and well, since what has value are not concrete individuals but diffuse systems or communities of life—or, indeed, the entire biosphere. By utilizing individual animals in certain ways we transcend the emotional bonds—the bonds of care and compassion for the individual—that would otherwise stand in the way to our achieving a higher sense of oneness with the community of all life. To the extent that an ecofeminist ethic must be rooted in an ethic of care, and assuming that such an ethic must find its expression in the caring relationships one has with *concrete individuals,* the ecstatic absorption of "self-in-Self" that is inseparable from deep ecology and the resultant devaluing both of discrete, concrete individuals, and of our protective care of them, must be anathema to ecofeminism.

The same is no less true of the vestiges of power and domination deep ecology sanctions, at least in the hunt. Here ecofeminists will deny that the *desire* to spill the "orgiastic blood" of another animal is "bred into the very fibre of the *human* race" or that it is "a deep and permanent yearning in the *human* condition." What it is, is a symptom of a virulent patriarchal sickness. When all the dust settles nature still must be subdued by the deep ecologist before it can be loved (assuming the two—subjugation and love—are even separable on this view). Considered against the history of the identification of nature and women it should not be surprising that ecofeminists will (or at least they should) see in the subjugation of the one, the subjugation of the other. When Eaton sweetly reports that "the hunter loves the animal he kills," eco- (and other) feminists might well think, "Yes, just as the rapist loves the woman he rapes." A defensible feminism, in my view, must be opposed to sport or recreational hunting—and this for the same kinds of reasons that it must be opposed to the violent exploitation of women.

No Clear Winner

But even if it is true, as I think it is, that ecofeminism and deep ecology remain divided by a deep, unbridgeable patriarchal gulf, this by itself is no argument for the superiority of the philosophy of ecofeminism. Indeed, I think it is not accurate to speak of ecofeminism as a "philosophy," if what we mean by this is a systematically developed position. Instead I think what ecofeminism offers us, at least at this point in time, is a series of provocative, possibly even correct insights into how we have come to think of ourselves in the ways we have and what some of the dangers of this way of thinking are. But how these insights are to be developed into a full-blown philosophy remains unclear. Ecofeminism, I think, is in a period of philosophical gestation, with the expected date of delivery still some time off. And there *will* be problems with the birthing. Allow me to indicate what some of these problems are and how I think they might be, if not solved, at least approached.

Some Problems for Ecofeminism

To begin with, ecofeminists must address the problem of how an ethic of care can overcome anthropocentric limits—how, that is, nurturing-care gets extended to wolves and coyotes, deer and whales, for example, not just family and friends. This problem is not solved by fixing the blame for our culture's thoughtless exploitation of wildlife or of nature generally on patriarchal attitudes and traditions. Even assuming that the blame belongs there, this finding still leaves us with the same question, namely, How can an ethic of care be extended beyond the boundaries of our own species?

Ecofeminists might take a page from the deep ecologist's script and answer that the way to enlarge the circle of compassionate care is by having certain experiences or, more generally, by experiencing the world in a certain way—experiencing nature as "thou," as it were, rather than as separate, alienated "it." And it is true that ecofeminists do stress the experiential aspects of their point of view ("a new consciousness for all of life," in Marti Kheel's words).[26] But the grave difficulty with this approach, if left unchecked by other considerations, is that experience (even widely shared experience) does not wear its veridical status on its sleeve, is not self-authenticating. One has only to recall the "consciousness of life" we find among racists to understand that experiencing-it-as-so does not make-it-so. Even if there are certain experiences that will open the heart (so to speak) to greater compassion, even if there is a way of experiencing the world from the compassionate point of view, the epistemological and moral superiority of either or both ideas remain to be authenticated.

But if not by experience alone, then how can an ethic of care overcome its tendency toward narrowness—toward, that is, caring for a few rather than the many? The approach I recommend involves reliance on appeals to consistency arguments.[27] By way of example: I care about my children's pain and suffering, not just because they are my children, but because of what pain and suffering are. This recognition of the locus of my care—it is their *being in pain* or their *suffering* that I care about—should lead me, if I am logical, to recognize the inappropriateness of my not caring about my neighbor's children's pain or suffering, or of my not caring about some stranger's children's pain or suffering, or, indeed, the inappropriateness of my not caring about the pain or suffering of any sentient being, human or otherwise. Whether I care or not, I *ought* to. And how I know that I ought to care is by recognizing that logic leads me to this conclusion. Of course, I might "see" this and yet not "feel" it. Respect for logic by itself, in isolation from everything else, does not create particular moral emotions or attitudes. What respect for logic can do in this case is help us recognize the arbitrariness of some of our emotions and the appropriateness of others.

It may do more. For I believe that this sort of cognitive realization, when it recurs over time, *can* give birth to the appropriate emotion. I do not say that this *always* happens. I only say that I think it sometimes can (in fact, I think it *has* happened in my own life). This is why I believe that, in this sense and for these reasons, it is not true to claim, as Hume does, that reason is, and *always* will be, "the slave of the passions." On the contrary, occasions can and do arise when our passions must be assessed and can be directed by our reason.

How much this is true can be demonstrated when we turn to consider where respect for reason must lead us in the case of our attitudes toward and treatment of sentient nonhuman animals. Most people are highly selective in the scope of their care, bestowing it on only comparatively few animals—companion animals and rare or majestic ones, for example. But pain is pain wheresoever it occurs, and an untimely death, when it is avoidable, is a great harm to any biographical being, whether human or not. Thus it is that our reason should lead us to recognize the arbitrariness of those feelings we have been culturally conditioned to have, feelings that cut us off emotionally from a sense of living-in-community with other sentient animals. The feelings of indifference many people have toward the avoidable suffering and untimely death of these animals are morally regrettable. They have been inculcated in us as a result of the two powerful forces mentioned earlier: namely, anthropocentrism and patriarchy. It is our reason that offers the means, and the hope, of our escape.

For the plain fact is, *billions* of nonhuman animals annually are caused gratuitous pain and suffer an untimely death, all for the sake of culturally conditioned human desires regarding food. These animals are in *desperate* need of our protection. For them it is *literally* a matter of life and death. Whether we speak of the moral ties that should bind us to them in the language of individual rights, as I am inclined to do, or in some other vocabulary, the essential practical point that must be made is that *we bear responsibility for what is happening to them*—there is blood on all our hands. That those in the camp of deep ecology fail to see the blood on their hands and instead delight in the "sport" of hunting shows, I believe, how radically deficient that philosophy is. None is so blind as he or she who will not see.

As should be evident, my earlier appeal to logic is an appeal to reason, to impartiality, to objectivity, to "the mind." In these respects some feminists will insist that my appeal bears the marks of patriarchal values (of "male mind"), and this may be enough to condemn it in the eyes of these thinkers. But the larger context in which this appeal is made concerns how the feminist ethic of care is to be extended beyond various arbitrary boundaries, a problem this ethic cannot avoid and a problem in the face of which, I believe, there can be no adequate solution if one abjures the use of reason. Thus, my appeal to reason is made, not as a means of repudiating what is false in feminism, but in an effort to understand what I think is true in it: the truth (in part) that emotion, not reason alone, is part of a full moral life, a truth that patriarchal moral theories conceal or deny.

8 Christians Are What Christians Eat

In its simplest terms the animal rights position I uphold maintains that such diverse practices as the use of animals in science, sport and recreational hunting, the trapping of fur-bearing animals for vanity products, and the practice of raising animals for human consumption are wrong because they systematically violate the rights of the animals involved. Morally, these practices ought to be abolished. This is the goal of the *social* struggle for animal rights. The goal of our *individual* struggle is to divest ourselves of our moral and economic ties to these injustices—for example, by not wearing the dead skins of animals and by not eating their decaying corpses.

Not a few people regard the animal rights position as extreme, calling, as it does, for the abolition of certain well-entrenched social customs rather than for their humane reform. And many seem to imagine that once this label ("extreme") has been applied, the need for further refutation evaporates. After all, how can an "extreme" moral position be correct?

I addressed this question in a recent speech, reminding my audience of a few extreme moral positions we all accept:

Rape is *always* wrong.
Child pornography is *always* wrong.
Racial and sexual discrimination are *always* wrong.

I went on to note that when an injustice is absolute, one must oppose it absolutely. It is not a "more humane" child pornography that an enlightened ethic calls for; it is its abolition that is required—it is this *extreme* position we must uphold. And similar remarks apply to the other examples.

Once this much is acknowledged it is evident (or at least it should be) that those who oppose or resist the animal rights position will have to do better than merely attach the label "extreme" to it. Sometimes extreme positions about what is wrong are right.

A Question of Justification

Of course there are two obvious differences between the animal rights position and these other extreme views. The latter are very generally accepted whereas the former is not. And unlike these very generally accepted views, which concern wrongful acts done to human beings, the animal rights position concerns wrongs done to animals (nonhuman animals, that is). Those who oppose or resist the animal rights position might seize upon these differences in an effort to justify themselves in accepting extreme positions regarding rape and child abuse, for example, while rejecting the "extremism" of animal rights.

But neither of these differences will bear the weight of justification. That a view is very generally accepted does not make it true. Time was when the shape of the earth was very generally believed to be flat, and time was when the presence of physical and mental handicaps were very generally thought to make the people who bore them morally inferior. That very many people believed these falsehoods did not turn them into truths. We do not discover or confirm what is true by taking a vote.

The reverse also can be demonstrated. That a view is not generally accepted does not prove it false. When only a few believed that the earth is round, and that the handicapped are the moral equals of others, they thought truly notwithstanding how grandly they were outnumbered. The solitary person who, in Thoreau's enduring image, marches to a different drummer may be the only person who apprehends the truth.

The second difference noted above is more problematic. Child abuse and rape involve evils done to human beings; the animal rights position claims that certain evils are done to nonhuman animals. This does constitute a difference. The question is, Is this a morally relevant difference—one that justifies extreme opposition to child abuse and rape but a less extreme opposition to vivisection and commercial trapping? For reasons I shall now explain, I do not believe so.

To begin with, this second difference only states certain biological facts: The victims of rape and child abuse belong to one species (the species *Homo sapiens*), whereas the victims of vivisection and trapping belong to other species (the species *Canis lupus,* for example). But biological differences *within* the species *Homo sapiens* (for example, in terms of sex, or skin color, or chromosome count) do not justify radically different treatment of those who differ. Why, then, should biological differences *outside* our species do so? If having one eye or deformed limbs does not disqualify a human being from moral consideration equal to that given to those humans who are more fortunate, how can it be rational to disqualify a rat or a wolf from equal moral consideration because, unlike us, they have paws and a tail?

Some of those who resist or oppose the animal rights position might have recourse to "intuition" at this point. They might claim that one either "sees" that the biological difference at issue (namely, species membership) *is* a morally relevant one, or one does *not* see this. The difference is "self-evident." Or so it will be claimed by those who "intuit" it.

However attractive this appeal to intuition may seem to some, it fails to bear the weight of justification. The plain fact is, people have claimed to "intuit" differences in the comparative moral standing of individuals and groups *inside* the human species. Over the course of history many men have "intuited" the moral superiority of men when compared with women, and many white-skinned humans have "intuited" the moral superiority of white-skinned humans when compared with humans having different skin colors. If this is a matter of intuition, then either one "sees" it or one does not. It's just that those who "see" it (or so they will insist) apprehend the truth, while those who fail to "see" it have deficient intuitive faculties.

I cannot believe that any thoughtful person will be taken in by this ruse. Appeals to "intuition" in these contexts are symptomatic of unquestioned and unjustifiable moral prejudice. What prompts or encourages men to "see" the moral inferiority of women are the sexual prejudices these men bring with them, not what is to be found in the existence of sexual differences. And the same is true, mutatis mutandis, of "seeing" moral superiority in racial or other biological differences between humans. In short, appeals to intuition, when made inside our species, and when they purport to discover the moral superiority latent within existing biological differences—such appeals have no role to play before the court of fair judgment.

This much established, the weakness of appeals to intuition in the case at hand should be apparent. Since intuition is not to be trusted when questions of the comparative moral worth of biologically different individuals *inside* the species *Homo sapiens* are at issue, it cannot be rational to assume that such appeals should be trusted when questions of the comparative moral worth of individuals *outside* our species are at issue. Moreover, since appeals to intuition in the former case turn out to be symptomatic of unquestioned and unjustifiable moral prejudice, it is not unreasonable to suspect that this same diagnosis applies in the case of appeals to intuition in the latter one. If true, then those who "intuit" the moral superiority of the species *Homo sapiens* over every other species also are in the grip of an unquestioned and unjustified moral prejudice.

"Speciesism" is the name commonly given to this prejudice, while those who believe and act on it are said to be "speciesists." Speciesists are people who think that humans and other animals should be treated differently because humans are humans and other animals are not. It would be wrong, they believe, to raise and slaughter humans for food but not wrong to do this to hogs and cows; wrong to create "models" of killer diseases in humans but not wrong to do the same thing to chimpanzees and rodents.

The prejudicial character of this belief is evident after appeals to "intuition" have been unmasked and after first one, then another unsuccessful attempt to prove human superiority has been defeated. Standard secular "proofs" fail miserably. Some (but not all) humans are rational. Some (but not all) humans are autonomous. Some (but not all) humans can use a language. Unless speciesists are prepared (which they are not) to consign humans who are deficient in these respects to the slaughterhouse or the research lab, they perforce must find some other characteristic possessed by all and only human beings.

Christianity and Speciesism

For those of us who cut our spiritual teeth on Christianity this challenge has a familiar response. Human beings—all of us—are said to be inherently more valuable than everything else because we are spiritually unique. This uniqueness stems from our having been created in the image of God. Thus, since all humans uniquely image God, we are able to cite a real (spiritual) difference between every member of our species and the millions of other species of creaturely life. And since this difference is a morally relevant one, Christian speciesists are able to defend their belief in human superiority. Or so it has been thought.

Now I am not ill-disposed to the idea that humans have a unique spiritual worth, nor am I ill-disposed to the idea that we possess it because we uniquely image God. However, the interpretation of these ideas I favor does not yield anything like the results favored by speciesism. This interpretation pictures humans as God's imaged presence in creation. By this I mean that we are expressly chosen by God to be God's vice-regent in the day-to-day affairs of the world; we are chosen by God, that is, to be as loving in our day-to-day dealings with the created order as God was in creating that order in the first place.

In *this* sense, therefore, there *is* a morally relevant difference between human beings and every other creaturely expression of

God. *Humans and humans alone* are given the awesome freedom and responsibility to be God's representative within creation. And it is, therefore, only we humans who can be held morally blameworthy when we fail to do this, and morally praiseworthy when we succeed.

With this interpretation of our unique imaging of God we find a morally relevant difference others have sought. But, as should be evident, this difference *by itself* offers neither aid nor comfort to the speciesist. For to agree that only humans image God, in the sense that only humans have the moral responsibility to be loving toward God's creation, does not entail *either* that all and only humans have inherent worth *or* that all and only humans have superior inherent worth. It is perfectly consistent with our unique status as God's chosen representative within creation that *other* creatures have inherent worth equal to our own.

When viewed within a Christian perspective, is this possibility actually true? In particular, how does the Bible speak to this question? As very much a nonexpert in the area of biblical exegesis I am reluctant to give an answer. But like the proverbial fool who "rushes in" I shall make bold and hazard the opinion that there is no one, unambiguous, unwaivering biblical message. Many passages lend support to viewing all of nonhuman creation as having no or little value apart from human needs and interests, a reading that supports the tradition of Christian anthropocentrism. By contrast, other passages support views that are more or less nonanthropocentric. I do not profess to know how to prove that the anthropocentric reading is false or that a nonanthropocentric reading is true. Indeed, as I already have indicated, I do not believe the Bible offers just one answer to our question.

The upshot, then, is that we are left with the awesome responsibility of choosing between alternative biblical representations of the value of nonhuman creation, no one of which is clearly or incontrovertably the correct one. And this fact should, I believe, chasten us in our conviction that we have privileged access to the whole truth, and nothing but the truth. With minds so feeble, spirits so weak, and a biblical message so open to honest differences of interpretation between people of real faith and goodwill, all who take spiritual sustenance from the pages of the Bible ought to realize both the need for, and the call to practice, the virtue of tolerance.

Eden's Message

Having said this, I may now speak to my own reading of the biblical message and indicate why this message, as I understand it, not only fails to offer aid and comfort to speciesists, it actually presents a healthy spiritual antidote to this virulent moral prejudice.

I take the opening account of creation in Genesis seriously, but not, I hasten to add, literally. I take it seriously because I believe this is the point from which our spiritual understanding of God's plans in and hopes for creation must begin, and against which our well-considered judgments about the value of creation finally must be tested. It is therefore predictable that I find it significant that God is said to judge each part of creation "good" before humans came upon the scene and that humans were created by God (or came upon the scene) on the same day as the nonhuman animals to whom I have been referring—those whose limbs are severed, whose sensory organs are brutally removed, and whose brains are ground up for purposes of scientific research, for example. I read in this representation of the order of creation a prescient recognition of the vital kinship humans share with these other animals, a kinship I have elsewhere endeavored to explicate in terms of our shared *biographical* presence in the world, a view which, quite apart from anything the Bible teaches, is supported by both common sense and our best science.

But I find in the opening saga of creation an even deeper, more profound message regarding God's plans in and hopes for creation. For I find in this account the unmistakable message that

God did *not* create nonhuman animals for our use—not in science, not for the purpose of vanity products, not for our entertainment, not for our sport or recreation, not even for our bodily sustenance. On the contrary, the nonhuman animals currently exploited in these ways were created to be just what they are: *independently good* expressions of the divine love that, in ways that are likely always to remain to some degree mysterious to us, was expressed in God's creative activity.

The issue of bodily sustenance is perhaps the most noteworthy of the practices I have mentioned since, while humans from "the beginning" were in need of food, there were no rodeos or circuses, no leghold traps or dynamite harpoons in the original creation. Had it been part of God's hopes in and plans for creation to have humans use nonhuman animals as food, it would have been open to God to let this be known. And yet what we find in the opening saga of creation is just the opposite. The "meat" we are given by God is not the flesh of animals, it is "all plants that bear seed everywhere on the earth, and every tree bearing fruit which yields seed: they shall be yours for food" (Gen. 1:29 [NEB]).

The message could not be any clearer. In the most perfect state of creation humans are *vegans* (that is, not only is the flesh of animals excluded from the menu God provides for us, even animal products—milk and cheese, for example—are excluded). And so I believe that, if we look to the biblical account of "the beginning" as more than merely one among many considerations, but instead as an absolutely essential source of spiritual insight into God's hopes in and plans for creation, then, like it or not, we are obliged to find there a menu of divinely approved bodily sustenance that differs quite markedly from the steaks and chops, the roasts and stews most Christians are accustomed to devouring.

To a less than optimal or scholarly degree I am aware of some of the chapters and verses of the subsequent biblical record: the Fall, the expulsion from the Garden, the Flood. There is no debate about the details of this record I could win if paired against an even modestly astute and retentive young person preparing for first communion. I wear my biblical (and theological) ignorance on my sleeve. Nevertheless, I believe the essential moral and spiritual truth any open-minded, literate reader of the first chapter of Genesis must find is the one I already have mentioned, namely, that nonhuman animals were not placed within God's creation so that humans might roast, fry, stew, broil, bake, and barbeque their rotting remains.

In this reading of God's creative activity, therefore, I find a spiritual lesson that is unmistakably at odds with both the letter and the spirit of speciesism. This lesson, as I understand it, does not represent the nonhuman animals to whom I have been referring as having no or less inherent value than humans. On the contrary, by unmistakably excluding these animals from the menu of food freely available to us, as granted by God's beneficence, I infer that God asks us to recognize the independent value of these animals. *They are not put here to be utilized by us. Rather, we are put here to protect them,* a message that is celestial music to the ears of one who, like myself, is not embarrassed or silenced by the "extremism" of the animal rights position. In this respect the (supposed) Christian defense of speciesism emerges as part of the best offense of the animal rights position.

Some theologians have a different view. Eden never was, they say; the perfection of creation is something we are to work to bring about in the future, not something that was lost in the past. I do not know how to prove which vision of Eden, if either, is the true one. What I do know is that, in the present context, this question is entirely moot, since it is clear—clear beyond any doubt, as I read the Scriptures—that human beings simply do not eat nonhuman animals in that fullness of God's beneficence the image of Eden represents. And this is true whether Eden once was (but was lost), or is yet to be (if we will but create it).

Every prejudice dies hard. Speciesism is no exception. That it is a prejudice and that, by acting on it, we humans have been, and continue to be, responsible for an incalculable amount of evil, an amount of truly monumental proportions, is as true as it is regrettable. In my philosophical writings over the past fifteen years I have endeavored to show how this tragic truth can be argued for on wholly secular grounds. On this occasion I have looked elsewhere for support, have in fact looked to the original saga of creation we find in Genesis in the hope that we might there find a religious or theological basis for animal rights, one that resonates with the secular case. Neither case—not the secular and not the religious—can have the conclusiveness of a proof in geometry. I say "can have" because I am reminded of Aristotle's astute observation when he writes that it is the mark of an educated person not to demand "proof" that is inappropriate for a given subject matter. Whatever else we might think of moral thought, we at least can agree that it is importantly unlike geometry.

Two Objections, Two Replies

It remains true, nonetheless, that my attempt to explain and defend an egalitarian view of the inherent value of humans and other animals must face a number of important challenges. For reasons of length, if for no other, I cannot respond to all of them, not even all the most fundamental ones. The best I can do, before concluding, is describe and defuse two of the most common objections.

The first begins by observing that, within the traditions of Judaism and Christianity, *every form of life,* not simply humans and other animals, is to be viewed as expressive of God's love. Thus, to attempt to "elevate" the value of nonhuman animals, as I may be accused of doing, is possible only at the expense of "lowering" the value of everything else, which is unacceptable.

I think this objection misses the mark. There is nothing in the animal rights philosophy that either denies or diminishes the value of fruits, grains, nuts, and other forms of vegetative life, or that refuses to accept the possibility that these life forms and the rest of creation generally are so many ways in which God's loving presence is manifested. Nor is there anything in this philosophy that disparages the wise counsel to treat all of creation gently and appreciatively. It is an arrogant anthropocentrism, often aided and abetted in our history by a no less arrogant Christian theology, not the philosophy of animal rights, that has brought the earth to the brink of ecological disaster.

Still, this philosophy does find in humans and other animals, because of our shared biographical status in creation, a kind of value—inherent value—that other creatures do not possess, either not at all or less than we do. Is it possible to defend this view? I believe it is, both on the basis of a purely secular philosophy and by appeal to biblical authority. The secular defense I have offered elsewhere and will not repeat here. As for the Christian defense, I shall merely reaffirm the vital importance of Genesis 1, which teaches, I believe, that vegetative life, not animal life, was meant to be used by us as food.

So much for the first challenge. The second one emanates from a quite different source and mounts a quite different objection. It begins by noting the large disparities that exist in the quality of life available to those who are affluent (the "haves") and those who are poor (the "have-nots"), especially those who live in Third World countries. "It is all well and good to preach the gospel of animal rights to those people who have the financial and other means to live it, if they choose to do so," this objection begins, "but please do spare us your self-righteous condemnation of the struggling—and often starving—masses of people in the world who really have no choice about what to eat or what to wear. To blame these people because they utilize animals is scandalous. It is to value animal life above human life. And this is misanthropy at its worst."

Now, this particular variation on the familiar theme of misanthropy (at least this is familiar to advocates of animal rights) has a point, up to a point. The point it has, is that *it would be* self-righteous to condemn the people in question for acting as they do, especially if we are acting worse than they are, as well we may. But, of course, nothing in what I have argued supports such a condemnation, and this for the simple reason that I have nowhere argued that people who eat animals, or who hunt and trap them, or who cut their heads off or burst their internal organs in pursuit of scientific knowledge, either are or must be evil people. The position I have set forth concerns the *moral wrongness* of what people do, not the comparative vileness of their character. In my view, it is entirely possible that good people sometimes do what is wrong, and evil people sometimes do what is right.

Indeed, not only is this possible, it frequently happens, and among those circumstances in which it does, some concern the actions performed by people in the Third World. To make my meaning clearer, consider the following hypothetical example. Suppose we chance upon a tribe of hunter–gatherers who annually, on a date sacred to their tradition, sacrifice the most beautiful female child to the gods, in the hope that the tribe will prosper in the coming year. In my view this act of human sacrifice is wrong and ought to be stopped (which does not mean that we should invade with tanks and flamethrowers to stop it!). From this moral assessment of the wrongness of what these people do, however, it does not follow that we should condemn them as evil, vicious people. For it could be that they act from only the best intentions and with nothing but the best motives. Nevertheless, what these people do, in my view, is wrong.

What is true of the imaginary case of this tribe, is no less true of real-life cases where people in the Third World raise and kill animals for food, cruelly subject other animals to forced labor, and so on. Anytime anyone reduces the inherent value of a nonhuman animal to that animal's utility value for human beings, what is done, in my view, is morally wrong. But it does not follow from this that we should make a negative judgment about the character of the human moral agents involved, especially if, as is true in the Third World, there are mitigating circumstances. For it often happens that people who do what is morally wrong should be *excused* from moral blame and censure. A person who injures a family member, for example, in the mistaken belief that there is a burglar in the house, does what is wrong and yet may not be morally blameworthy. Similarly, those people in the Third World who act in ways that are wrong according to the philosophy of animal rights, do what is wrong. But because of the harsh, uncompromising exigencies of their life, where it often literally is a matter of their life or their death that hangs in the balance, the people in the Third World in my view should be excused from our harsh judgment of moral blame. The circumstances of their lives, one might say, are as mitigating as any circumstances can be.

In light of the preceding I hope it is clear why it would be a bad reading of the philosophy of animal rights to charge its proponents with a hearty appetite, if not for animal flesh, then at least for self-righteousness. When we understand the differences between morally assessing a person's act and morally assessing that person's character, and when we take cognizance of the appropriateness of reducing or eliminating moral blame in the face of mitigating circumstances, then the proponents of animal rights should be seen to be no more self-righteous than the proponents of any other philosophy.

Conclusion

Finally, then, in closing, I wish to make a few observations closer to home, so to speak. Most of us who are in attendance here have traveled hundreds or thousands of miles at the cost of irreplaceable fuels, the production and combustion of which contribute to the pollution of air and water, and to the deforestation of the earth's woodlands. We are housed comfortably, are the beneficiaries of indoor plumbing and hot showers, eat wholesome, tasty food—and all this while the great majority of our fellow humans scrape by, catch-as-catch-can, from one day to the next. And we have journeyed here, and are gathered together here, leisurely to discuss issues relating to the integrity of creation. Truly we are among the lucky ones—the sons and daughters of a capricious privilege.

Just as surely, in my view, we daily run the risk of succumbing to a detached hypocrisy. For the questions we must face concern not only the *idea* of the integrity of creation, they also ask how we *should live* if we are to express our thoughtful allegiance to this idea in our day-to-day life. This ancient question has no simple answer. There is much good that we would do, that we do not. And much evil that we would not do, that we find ourselves doing. The challenge to lead a good, respectful, loving life, just in our dealings with members of the human family, is onerous. How much more demanding will it be, then, if we widen the circle of the moral community to include the whole of creation.

How might we begin to meet this larger challenge? Doubtless there are many places to begin, some of which will be more accessible to some than to others. For my part, however, I cannot help believing that an appropriate place to begin is with the food on our plates. For here we are faced with a direct personal choice over which we exercise absolute sovereign authority. Such power is not always within our grasp. How little influence each of us really has when it comes to the practices of the World Bank, the agrarian land-reform movement, the cessation of armed conflict, crime, drug use, the scourge of famine. These large-scale evils stand beyond the reach of our small wills.

But not the food on our plates. Here we are at liberty to exercise absolute control. And here, then, it is only reasonable that we ask ourselves whether the choices we make about the food we eat have any bearing on our thoughtful attempt to honor the integrity of creation.

When we consider the biographical and, I dare say, the spiritual kinship we share with those billions of animals annually raised and slaughtered to be eaten; when, further, we inform ourselves of the truly wretched conditions in which most of these animals are raised, not to mention the deplorable methods by which they are transported and the gruesome, blood-soaked reality of the slaughterhouse; when we apprise ourselves of the massive destruction of the earth's ecology that, both directly and indirectly, is attributable to commercial animal agriculture; and when, finally, we take honest stock of our privileged position in the world, a position that will not permit us the excuse from moral blame shared by the desperately poor who, as we say, really have no choice—when we consider all these factors, then the case for abstaining from animal flesh has the overwhelming weight of both impartial reason and a spiritually infused compassion on its side.

True, to make this change will involve some sacrifices—in taste perhaps, in convenience certainly. And yet the whole fabric of Christian agape is woven from the threads of sacrificial acts. To abstain, on principle, from eating other animals, therefore, although it is not the end-all, can be the begin-all of our conscientious effort to journey back (or forward) to Eden, can be one way among others in which we reestablish or create that relationship to the earth that, if Genesis 1 is to be trusted, was part of God's original hopes for and plans in creation. It is the integrity of this creation we seek to understand and aspire to honor. In the choice of our food, I believe, we see, not through a glass darkly, but face-to-face, a small but not unimportant part of what must be done. For despite the prodigiousness of our ignorance, we do know this: Christians are what Christians eat.

Notes and Readings

Chapter 1

In the fall of 1988 Peter French, the new editor of the *Journal of Social Philosophy,* asked me if I would contribute an essay to a special issue of that quarterly devoted to the future of philosophy. Contributors were encouraged to let their imaginations run free and to play the unusual role of the "visionary." The result of my peering-into-the-future is the short piece that, with minor changes, is reproduced here. "The Thee Generation" was originally published in the *Journal of Social Philosophy* 20, nos. 1, 2 (Spring/Fall 1989): 31–33, and is reprinted here with permission of the journal and its editor.

For Further Reading

All of the references listed in the pages that follow, both in the notes and in the suggestions for further reading, bear on the larger themes addressed in "The Thee Generation." All, therefore, are recommended for further study, the following in particular.

Ehrenfield, David. *The Arrogance of Humanism.* New York: Oxford University Press, 1978.

Gilligan, Carol. *In a Different Voice.* Cambridge, Mass.: Harvard University Press, 1982.

Hartshorne, Charles. *Beyond Humanism.* Lincoln: University of Nebraska Press, 1968.

Regan, Tom. *The Struggle for Animal Rights.* Clarks Summit, Pa.: International Society for Animal Rights, 1987.

Chapter 2

In November 1988 I was invited to deliver the keynote address for a series of lectures entitled Ethics and the Professions, sponsored by the Florence Hecht Residential College of the University of Miami in Florida. Earlier that same year, in August, I had been invited to present a paper at the annual meeting of the American Psychological Association on the general theme "Perpetuating Injustice: Social Perception and Exclusion from the Moral Community." The essay printed here incorporates parts of both of these lectures. In it I sketch the methods and objectives of moral philosophy, as I understand this discipline. My understanding of these and other matters has been influenced by many people but none more so than James Rachels, to whom I owe a special debt. In addition to some remarks about its methods and objectives, I also say something about the motivation for doing moral philosophy and relate my findings to a particular question which, in the past, most people have failed to ask when deliberating about what profession they might pursue. This question concerns the moral status of nonhuman animals. Members of The Thee Generation *do* ask this question—and examine it thoughtfully—before they decide on their careers. This is one way in which members of this generation differ from others. I offer my own answer to the question about our duties to other animals in Chapter 3, "Ill-Gotten Gains."

Note

1. The quotation from Einstein is from a letter published in the *New York Post,* November 28, 1972.

For Further Reading

Care, Norman S. *On Sharing Fate.* Philadelphia: Temple University Press, 1987. See, especially, chap. 2, "Career Choice," and chap. 8, "On the Importance of One's Life."

Goodpaster, Kenneth, "On Being Morally Considerable." *Journal of Philosophy* 75 (1978): 308–25.

Nagel, Thomas. *Mortal Questions.* Cambridge, England: Cambridge University Press, 1979. See esp. "Moral Luck."

Rachels, James. *The Elements of Moral Philosophy.* Philadelphia: Temple University Press (cloth); New York: McGraw-Hill (paper), 1988.

Regan, Tom, and Peter Singer, eds. *Animal Rights and Human Obligations.* 2d ed. Englewood Cliffs, N.J.: Prentice-Hall, 1989.

Singer, Peter. *Animal Liberation.* 2d ed. New York: Random House, 1990.

———. "Animals and the Value of Life." In *Matters of Life and Death,* 2d ed., edited by Tom Regan. New York: McGraw-Hill, 1985.

Wenz, Peter. *Environmental Justice.* Albany: State University of New York Press, 1987. See esp. chap. 15, "The Individual's Obligations."

Williams, Bernard. *Moral Luck.* Cambridge, England: Cambridge University Press, 1981.

Chapter 3

In this essay I explore the question posed at the end of Chapter 2, namely, Morally, what do we owe the other animals? Biomedical research is the particular context in which this general question is examined. I argue that it is morally indefensible to use other animals in this kind of research, which is why it must be stopped, and stopped completely. This position may seem "extreme" to some (perhaps even most) people. But one of the characteristics that defines The Thee Generation is a shared belief in a point established in Chapter 2: Unexamined moral beliefs, including those favored by the majority, are not necessarily the correct ones. To find the truth we must venture outside the cave of conventional morality. This is what I attempt to do in this chapter. That the position I bring back differs from the one most people accept is therefore hardly surprising. This chapter is a somewhat revised version of an essay that originally appeared in *Health Care Ethics,* edited by Donald Van DeVeer and Tom Regan (Philadelphia: Temple University Press, 1987), pp. 239–64. It is reprinted here with the permission of the publisher. For a more systematic defense of my position, see my *The Case for Animal Rights* (Berkeley: University of California Press; London: Routledge and Kegan Paul, 1983).

Notes

1. Cora Diamond, "Experimenting on Animals: A Problem in Ethics," in *Animals in Research: New Perspectives in Animal Experimentation,* ed. David Sperlinger (New York: John Wiley and Sons, 1981), p. 345. The bibliography to which Diamond refers is S. Sollitto and R. M. Veatch, eds., *Bibliography of Society, Ethics, and the Life Sciences, 1979–1980* (Hastings-on-Hudson, N.Y.: The Hastings Center, 1978).
2. Charles R. Magel, *A Bibliography on Animal Rights and Related Matters* (Washington, D.C.: University Press of America, 1981).
3. Diamond, "Experimenting on Animals," p. 345.

For Further Reading

Langley, Gil, ed. *Animal Experimentation: The Consensus Changes.* London: Macmillan, 1989.

Magel, Charles. *Keyguide to Information Sources in Animal Rights.* Jefferson, N.C.: McFarland, 1988.

Midgley, Mary. *Animals and Why They Matter.* Athens: University of Georgia Press, 1984.

Rachels, James. *Created from Animals: The Moral Implications of Darwinism.* Oxford and New York: Oxford University Press, 1990.

Regan, Tom. "The Case for Animal Rights." In *The Struggle for*

Animal Rights. Clarks Summit, Pa.: International Society for Animal Rights, 1987.

Regan, Tom, and Peter Singer, eds. *Animal Rights and Human Obligations.* 2d ed. Englewood Cliffs, N.J.: Prentice-Hall, 1989.

Rollins, Bernard, *The Unheeded Cry.* New York: Oxford University Press, 1989.

Rupke, Nicholaas A., ed. *Vivisection in Historical Perspective.* New York: Croom Helm, 1987.

Ryder, Richard. *Animal Revolution: Changing Attitudes Towards Speciesism.* Oxford: Blackwells, 1989.

———. *Victims of Science.* Fontwell, England: Centaur Press, 1983.

Sharpe, Robert. *The Cruel Deception: The Use of Animals in Medical Research.* London: Thorsons, 1988.

Singer, Peter. *Animal Liberation.* 2d ed. New York: Random House, 1990.

Chapter 4

The position sketched in Chapter 3, "Ill-Gotten Gains," might strike some people as "extreme," calling, as it does, for the total abolition of the harmful use of animals in science. It is therefore important to realize that the theory on which this antivivisectionist position rests is of considerable power when applied to vulnerable populations within the human family. This general point is illustrated in this chapter on child pornography. If sound, what this and Chapter 5, "Feminism and Vivisection" show is important: The most adequate philosophical account of why child pornography is wrong also happens to be the same account that shows why vivisection is wrong. More generally, what success in this quarter shows is that the The Thee Generation's ethic of service and protection cannot be well grounded without recognizing the value and rights of the individual, whether the individual is a human being (and, if human, whether an adult or a child), or some other animal. A longer version of "Why Child Pornography Is Wrong" was first published in *Children, Parents and Politics,* edited by G. Scarre (Cambridge, England: Cambridge University Press, 1989), pp. 182–204. This shorter version is produced here with the permission of the publisher.

Notes

1. I am well aware that this is not an uncontroversial assumption to make about the assessment of moral theories. I offer a defense of this assumption in my *The Case for Animal Rights* (Berkeley: University of California Press; London: Routledge and Kegan Paul, 1983), esp. chap. 4.

2. *New York v. Ferber,* 458 U.S. 474, 746 (1982). Cited in *Attorney General's Commission on Pornography: Final Report, July 1986* (Department of Justice, Washington, D.C.: U.S. Government Printing Office), pp. 596–97. My use of the Meese Commission's *Report* does not imply acceptance of all of its findings.

3. Ibid., p. 601.

4. Ibid., p. 611.

5. *New York v. Ferber,* 458 U.S. 474, at 756–57, *Commission on Pornography,* p. 658.

6. *Commission on Pornography,* p. 613.

7. Ibid., p. 609.

8. Ibid.

9. Jan Narveson on a number of occasions has argued for a contractarian position of this sort. See, for example, "At Arms' Length: Violence and War," in *Matters of Life and Death: New Introductory Essays in Moral Philosophy,* ed. Tom Regan, 2d ed. (New York: Random House, 1986), pp. 125–74.

10. See Rawls's seminal work, *A Theory of Justice* (Cambridge, Mass.: Harvard University Press, 1971).

11. I believe James Rachels was the first person to make this point in these words.

12. I have benefited from criticisms of an earlier draft of this essay raised by Matt Cartmill, Melinda Vadas, Donald Van DeVeer, and the members of the Society for Philosophy and Public Policy of New York. The views expressed and the approach to the issues taken here, however, are mine.

For Further Reading

Andre, Judith. "Poole on Obscenity and Censorship." *Ethics* 94 (1984): 496–500.

Beis, R. H. "Pornography and the Harm It Does." *International Journal of Morality and Social Science* 2 (1987): 81–92.

Brongersma, Edward. *Loving Boys: A Multidisciplinary Study of Sexual Relations Between Adult and Minor Males.* Elmhurst, N.Y.: Global Academic Publishers, 1986.

D'Amico, Robert. "The Meaning of Pornography." *Human Society* 7 (1984): 87–102.

Jarvie, I. C. "The Sociology of the Pornography Debate." *The Philosophy of Social Science* 17 (1987): 257–75.

Linton, David. "Why Is Pornography Offensive?" *Journal of Value Inquiry* 12 (1979): 57–62.

McHarry, Mark. "The Meese Commission: Ominous Implications for Gay Men and Lesbians." *Nambla Bulletin: Voice of the North American Man/Boy Love Association* 7 (1986): 11.

O'Brien, Shirley. *Child Pornography.* 2d ed. Dubuque, Iowa: Kendall-Hunt, 1983.

O'Connor, Tom. *Paedophilia: The Radical Case.* See, especially, chap. 10, "Children in Erotica and Pornography." London: P. Owen, 1982.

Poole, Howard. "Obscenity and Censorship." *Ethics* 93 (1982): 39–44.

Wendell, Susan. "Pornography and Freedom of Expression." In *Philosophy and Culture,* edited by Venant Cauchyy, vol. 1, pp. 236–40. Laval, P. Que., 1988.

Wilson, Glenn David, and David Cox. *Child Lovers: A Study of Paedophilia in Society.* London: P. Owen, 1983.

Chapter 5

In the spring of 1988 I was privileged to deliver the annual Ryle Lectures at Trent University, Peterborough, Ontario. Named in honor of the distinguished twentieth-century English philosopher Gilbert Ryle, these lectures are presented on four successive nights to an audience of both professionals and lay people, as well as undergraduates from throughout the university. The title of my lectures was "Reconsidering Individualism." In them I explored a variety of criticisms being lodged against the importance *the individual* has enjoyed in Western ethical theory in general and my own theory in particular. I was especially interested in learning more about feminist philosophy and why my moral theory had to face challenges stemming from this quarter. The essay printed here is the third of my Ryle lectures, one that I was again privileged to present as the Wayne A. R. Leys Memorial Lecture at Southern Illinois University–Carbondale, also in the spring of 1988. My particular interest in this essay is whether the antivivisectionist position I defend—for example, in Chapter 3, "Ill-Gotten Gains"—is at odds with one or another tendency in feminist thought. Because I believe that the philosophy of animal rights and a progressive feminist philosophy, as I understand this idea, share many of the aspirations of The Thee Generation, I hoped I would find agreement on the issue of vivisection. Whether I did find it, or instead put it there, is something thoughtful readers will have to decide for themselves.

Notes

1. Mary Wollstonecraft, *A Vindication of the Rights of Women* (London: Dent, 1982), p. 15.
2. Val Plumwood, "Women, Humanity and Nature," *Radical Philosophy* 48 (1988): 22.
3. Amelia Bloomer, "Response to Mr. T. S. Arthur's 'Ruling a Wife,'" quoted in *Feminist Quotations: Voices of Rebels, Re-*

formers, and Visionaries, ed. Carol McPhee and Ann Fitzgerald (New York: Crowell, 1979), p. 124.

4. Ariel Salleh, "Contributing to the Critique of Political Epistemology," *Thesis Eleven,* no. 8 (1984).

5. Plumwood, "Women, Humanity and Nature," p. 26.

6. Marilyn French, *Beyond Power: Of Women, Men and Morals* (New York: Ballantine Books, 1986), p. 341.

7. Carol Gilligan, *In a Different Voice: Psychological Theory and Women's Development* (Cambridge, Mass.: Harvard University Press, 1982).

8. Lynda Birke, *Women, Feminism and Biology: The Feminist Challenge* (London: Routledge, Chapman, and Hall, 1986), p. 159.

9. *Alternatives to Animal Use in Research, Testing and Education* (New York: Dekker, 1988), p. 214.

10. Birke, *Women, Feminism and Biology,* p. 159.

11. These words are part of a poem entitled "Somnia Medici" that appeared in *Zoophilist* 5 (1 May 1885): 1. The author is unknown.

12. Elie de Cyon, "The Anti-Vivisectionist Agitation," *Contemporary Review* 43 (1883): 506.

13. I am especially indebted to Carol Adams for her many helpful criticisms of the original draft of this chapter. Whatever errors remain are solely my responsibility.

For Further Reading

Birke, Lynda. " 'They're Worse Than Animals': Animals and Biological Research." In *More Than the Parts: Biology and Politics,* edited by Lynda Birke and John Silvertown, pp. 219–35. London: Pluto Press, 1984.

Blier, Ruth, ed. *Feminist Approaches to Science.* New York: Pergamon Press, 1986.

Easlea, B. *Science and Sexual Oppression: Patriarchy's Confrontation with Women and Nature.* London: Weidenfeld and Nicholson, 1981.

Elston, Mary Ann. "Women and Anti-Vivisection in Victorian England." In *Vivisection in Historical Perspective,* edited by Nicholaas A. Rupke, pp. 259–94. New York: Croom Helm, 1987.

Harding, Sandra. *The Science Question in Feminism.* Ithaca, N.Y.: Cornell University Press, 1986.

Keller, Evelyn Fox. *Reflections on Gender and Science.* New Haven, Conn.: Yale University Press, 1985.

Lloyd, Genevieve. *The Man of Reason: "Male" and "Female" in Western Philosophy.* Minneapolis: University of Minnesota Press, 1984.

Midgley, Mary. *Animals and Why They Matter.* Athens: University of Georgia Press, 1984.

Noddings, Nel. *Caring: A Feminist Approach to Ethics and Moral Education.* Berkeley: University of California Press, 1984.

Tuana, Nancy, ed. *Newsletter on Feminism and Philosophy,* November 1987. Newark, Del.: American Philosophical Association.

Chapter 6

In October 1982 I was asked to make a contribution to a multidisciplinary conference organized on the theme "Agriculture, Change and Human Values," sponsored by the Humanities and Agriculture Program at the University of Florida. Looking back on this occasion I do not hesitate to say that I took home more than I brought with me. Political scientists and economists, advocates of small-scale organic and of large-scale corporate farming, government policymakers and informed critics of such policy, representatives of minority farmers and of Third World agriculture: Just about every important voice was given an opportunity to be heard, even those who aspired to speak for those animals who cannot speak for themselves.

My original presentation has been revised extensively; it is more discursive, less argumentative than the original. I am not so much interested in proving that my views are right, those of others wrong, as I am in explaining how, despite their many differences, some of the philosophical partisans who make up The Thee Generation can agree on some things—and these, far from trivial things. In the world as we find it today, the diet of The Thee Generation is vegetarian. The major philosophies vying to lead the coming revolution all agree that the consumption of commercially raised farm animals, as these animals are raised today, must become a thing of the past. And it can become a thing of the past only if we individually and collectively become vegetarians. This is the main conclusion for which I argue in this chapter. Along the way, however, I also indicate some of the problems these competing philosophies must face.

Notes

1. Kant's views are set forth in a short section of his *Lectures on Ethics,* trans. Louis Infield (New York: Harper and Row, 1963), pp. 239–41. They are included in *Animal Rights and Human Obligations,* ed. Tom Regan and Peter Singer (Englewood Cliffs, N.J.: Prentice-Hall, 1976, 1st ed., pp. 122–23; and 1989, 2d ed., pp. 23–24).
2. Thomas Aquinas, *Summa Contra Gentiles,* literally translated by the English Dominican Fathers (Benziger Brothers, 1928), Book 3, pt. 2, chap. 112. For a more readily available source, see Regan and Singer, *Animal Rights and Human Obligations,* 1st ed., pp. 58–59; and 2d ed., pp. 8–9.
3. Jeremy Bentham, *The Principles of Morals and Legislation* (1789), chap. 17, sec. 1. Reprinted in Regan and Singer, *Animal Rights and Human Obligations,* 1st ed., p. 130; and 2d ed., p. 26.
4. John Stuart Mill, "Whewell on Moral Philosophy," in Mill, *Collected Works* (Toronto: University of Toronto Press, 1962–77), vol. 10, p. 187. Reprinted in Regan and Singer, *Animal Rights and Human Obligations,* 1st ed., p. 132.

5. See, in particular, Peter Singer, *Animal Liberation,* 2d ed. (New York: Random House, 1990).

6. Richard Ryder, *Speciesism: The Ethics of Vivisection* (Edinburgh: Scottish Society for the Prevention of Vivisection, 1974). See also his *Victims of Science* (Fontwell, England: Centaur Press, 1983).

7. See, in particular, Chapter 3, "Ill-Gotten Gains."

8. For my responses to the problems raised against the rights view, see *The Case for Animal Rights* (Berkeley: University of California Press; London: Routledge and Kegan Paul, 1983).

9. See, in particular, Aldo Leopold, *A Sand County Almanac* (New York: Baltimore Books, 1970).

10. For a compelling response to this view, see Ron Baker, *The American Hunting Myth* (New York: Vantage Press, 1987).

11. See, for example, J. Baird Callicott, "In Search of an Environmental Ethic," in *Matters of Life and Death,* 2d ed., ed. Tom Regan, pp. 381–424 (New York: McGraw-Hill, 1986).

12. The best account of these results is found in Jim Mason and Peter Singer, *Animal Factories* (New York: Crown, 1980).

13. Regan, *The Case for Animal Rights,* esp. chaps. 7 and 8.

For Further Reading

Adams, Carol J. *The Sexual Politics of Meat: A Feminist–Vegetarian Critical Theory.* New York: Continuum, 1990.

Aiken, William. "Ethical Issues in Agriculture." In *Earthbound: New Introductory Essays in Environmental Ethics,* edited by Tom Regan, pp. 247–88. Philadelphia: Temple University Press, 1984.

Akers, Keith. *A Vegetarian Sourcebook: The Nutrition, Ecology, and Ethics of a Natural Foods Diet.* Arlington, Va.: Vegetarian Press, 1989.

Berry, Wendell. *A Continuous Harmony: Essays Cultural and Agricultural.* New York: Harcourt Brace Jovanovich, 1970.

Dawkins, Marian. *Animal Suffering: The Science of Animal Welfare.* New York: Routledge, Chapman, and Hall, 1980.

Giehl, Dudley. *Vegetarianism: A Way of Life.* New York: Harper and Row, 1979.

Harrison, Ruth. *Animal Factories.* London: Vincent Stuart, 1964.

Jackson, Wes. *New Roots for Agriculture.* San Francisco: Friends of the Earth, 1980.

Lappé, Francis Moore. *Diet for a Small Planet.* New York: Ballantine Books, 10th anniversary edition, 1985.

Robbins, John. *Diet for a New America: How Your Food Choices Affect Your Health, Happiness and the Future Life on Earth.* Walpole, N.H.: Stillpoint, 1987.

Schell, Orville. *Modern Meat.* New York: Random House, 1984.

Wellford, Harrison. *Sowing the Wind.* New York: Grossman, 1972.

Wenz, Peter. *Environmental Justice.* Albany: State University of New York Press, 1988.

Wynne-Tyson, Jon. *Food for a Future: How World Hunger Could Be Ended by the Twenty-first Century,* rev. ed. Wellingborough, Northants: Thorsons, 1988.

Chapter 7

This chapter is the fourth and final of my Ryle Lectures (see also Chapter 5). It was also my good fortune to be invited to read this same essay at an international conference on environmental ethics at the Instituto Suor Orsola Bienincasa, in Naples, Italy, in July 1988. In it I examine the common ground shared by two of the major voices speaking in behalf of The Thee Generation: ecofeminists and deep ecologists. There are differences in what they say, and not only are these differences important, some seem to be irreconcilable. Whatever doubts I may have had on this matter have been quieted by the illuminating and persuasive work of Marti Kheel (see references, below). At a certain level, however, there are points of harmony among ecofeminists and deep ecologists, and it is this harmony that is nurturing The Thee Generation and the promise of moral revolution this generation carries with it.

Notes

1. Carolyn Merchant, *The Death of Nature: Women, Ecology, and the Scientific Revolution* (New York: Harper and Row, 1980), p. 100.
2. Ibid., pp. 99–100.
3. Ibid., p. 193.
4. Ibid., p. 2.
5. Ibid., pp. 6–7.
6. Ibid.
7. Ibid.
8. Ibid., p. 2.
9. Ibid., p. 165.
10. Ibid., p. 169.
11. Ibid., p. 172.
12. Ibid.
13. These ideas are discussed at greater length in the opening sections of Chapter 5, "Feminism and Vivisection."
14. Warwick Fox, "The Intuition of Deep Ecology," quoted in Bill Devall and George Sessions, *Deep Ecology: Living as if Nature Mattered* (Salt Lake City: Peregrine Smith Books, 1985), p. 66.
15. Ibid., pp. 66–67.
16. Ibid., pp. 67–68.
17. Ibid., p. 66.
18. Ibid., p. 69.
19. Ibid., p. 75.
20. Marti Kheel, "Ecofeminism and Deep Ecology: Reflections on Identity and Difference," forthcoming in *Reweaving the World: The Emergence of Eco-feminism,* ed. Irene Diamond and Gloria Orenstein (San Francisco: Sierra Club, in press). My debts to Kheel are both obvious and great. I am especially indebted to her for her compelling analysis of the prohunting literature authored by deep ecologists, to which I refer in what follows.
21. José Ortega y Gasset, *Meditations of Hunting,* trans. Howard B. Wescott, with a foreword by Paul Shepard (New York: Charles Scribner's Sons, 1985), p. 92.

22. Randall L. Eaton, "The Hunter as Alert Man: An Overview of the Origin of the Human/Animal Connection," in *The Human/Animal Connection,* ed. Randall L. Eaton (Incline Village, Nev.: Carnivore Journal and Sierra Nevada College Press, 1985), p. 9.

23. Ortega, *Meditations of Hunting,* p. 29.

24. Aldo Leopold, *A Sand County Almanac* (New York: Baltimore Books, 1970), p. 227.

25. Kheel, "Ecofeminism and Deep Ecology," p. 22 of typescript.

26. Ibid., p. 1.

27. See Chapter 5, "Feminism and Vivisection."

For Further Reading

Adams, Carol J. *The Sexual Politics of Meat: A Feminist–Vegetarian Critical Theory.* New York: Continuum, 1990.

Battin, Peter. *Living Trophies.* New York: Crowell, 1976.

Callicott, J. Baird. "Animal Liberation: A Triangular Affair." *Environmental Ethics* 2 (1980): 331–38.

———. *In Defense of the Land Ethic: Essays in Environmental Philosophy.* Albany: State University of New York Press, 1989.

Cheney, Jim. "Eco-Feminism and Deep Ecology." *Environmental Ethics* 9 (1987): 115–45.

Fox, Warwick. "Approaching Deep Ecology: A Response to Richard Sylvan's Critique of Deep Ecology." *Environmental Studies Occasional Papers* 20 (University of Tasmania, 1986).

Johnson, Edward. "Treating the Dirt: Environmental Ethics and Moral Theory." In *Earthbound: New Introductory Essays in Environmental Ethics,* edited by Tom Regan, pp. 335–66. Philadelphia: Temple University Press, 1984.

Kheel, Marti. "The Liberation of Nature: A Circular Affair." *Environmental Ethics* 7 (1985): 135–49.

Moline, Jon. "Aldo Leopold and the Moral Community." *Environmental Ethics* 8 (1986): 99–120.

Naess, Arne. "The Deep Ecological Movement: Some Philosophical Aspects." *Philosophical Inquiry* 8 (1986): 10–31.

———. "A Defense of the Deep Ecology Movement." *Environmental Ethics* 6 (1984): 265–70.

———. "The Shallow and the Deep, Long-Range Ecology Movements: A Summary." *Inquiry* 16 (1973): 95–100.

Nash, Roderick. *Wilderness and the American Mind.* 3d ed. New Haven, Conn.: Yale University Press, 1982.

Rolston, Holmes, III. *Environmental Ethics: Duties to and Values in the Natural World.* Philadelphia: Temple University Press, 1988.

———. *Philosophy Gone Wild.* Buffalo, N.Y.: Prometheus Books, 1986.

Salleh, Ariel Kay. "Deeper than Deep Ecology: The Eco-Feminist Connection." *Environmental Ethics* 6 (1984): 339–45.

Sylvan, Richard. "A Critique of Deep Ecology." *Radical Philosophy* 40 (1985): 2–12.

Warren, Karen J. "Feminism and Ecology: Making Connections." *Environmental Ethics* 9 (1987): 3–20.

Watson, Richard A. "A Critique of Anti-Anthropocentric Biocentrism." *Environmental Ethics* 5 (1983): 245–56.

Zimmerman, Michael E. "Feminism, Deep Ecology, and Environmental Ethics." *Environmental Ethics* 9 (1987): 21–44.

Chapter 8

In September 1988 I had the honor of being asked to make a presentation before a World Council of Churches' consultation on the theme "Justice, Peace, and the Integrity of Creation." This was a historic moment. Never before had the WCC considered the idea of animal rights on its merits, having previously addressed the ethics of our treatment of other animals only tangentially (for example, as this arises within the context of concern about endangered species). My challenge was to determine whether an informed, compassionate Christian ethic makes serious demands on how *plentiful* animals are to be treated. In order that my presentation might have some practical bite, so to speak, I concentrated primarily on dietary matters. Since it is true, in my view, that vegetarianism is the diet of The Thee Generation (see Chapter 6, "Abolishing Animal Agriculture), and since I hoped that a person's Christian beliefs would not bar that person from being a part of this generation, I wanted to see whether and, if so, how one might argue for this dietary way of life on distinctively Christian grounds. With few changes, this chapter represents my attempt to meet these challenges. For the original essay, see *Liberating Life: Contemporary Approaches to Ecological Theology,* edited by Jay McDaniel. Maryknoll, N.Y.: Orbis Press, 1990.

For Further Reading

Berry, Thomas. *The Dream of the Earth.* San Francisco: Sierra Club Books, 1988.

Birch, Charles, and John Cobb. *The Liberation of Life: From Cell to Community.* Cambridge, England: Cambridge University Press, 1981.

Clark, Stephen. *The Moral Status of Animals.* Oxford: Clarendon, 1977.

Dombrowski, Daniel. *Hartshorne and the Metaphysics of Animal Rights.* Albany: State University of New York Press, 1988.

Gustafson, James. *Ethics from a Theocentric Perspective.* Vol. 1, *Theology and Ethics.* Chicago: University of Chicago Press, 1981.

———. *Ethics from a Theocentric Perspective.* Vol. 2, *Ethics and Theology.* Chicago: University of Chicago Press, 1984.

Hall, Douglas. *Imaging God: Dominion as Stewardship.* Grand Rapids, Mich.: Eerdmans, 1986.

Hartshorne, Charles. *Beyond Humanism.* Lincoln: University of Nebraska Press, 1968.

———. "Foundations for a Human Ethic: What Human Beings Have in Common with Other Higher Animals." In *On the Fifth Day: Animal Rights and Human Ethics,* edited by Richard Knowles Morris and Michael W. Fox. Washington, D.C.: Acropolis Press, 1978.

Linzey, Andrew. *Animal Rights: A Christian Assessment of Man's Treatment of Animals.* London: SCM Press, 1976.

———. *Christianity and the Rights of Animals.* New York: Crossroad, 1987.

Linzey, Andrew, and Tom Regan, eds. *Animals and Christianity: A Book of Readings.* London: SPCK; New York: Crossroad, 1989.

McDaniel, Jay B. *Earth, Sky, Gods and Mortals.* Mystic, Conn.: Twenty-third Publishing Co., 1990.

———. *Of God and Pelicans: A Theology of Reverence for Life.* Louisville, Ky.: Westminster/John Knox Press, 1989.

Regan, Tom, ed. *Animal Sacrifices: Religious Perspectives on the Use of Animals in Science.* Philadelphia: Temple University Press, 1986.

Rolston, Holmes, III. *Science and Religion: A Critical Survey.* New York: Random House, 1987.

Schweitzer, Albert. *Reverence for Life,* edited by R. H. Fuller. London: SPCK, 1970.

Index

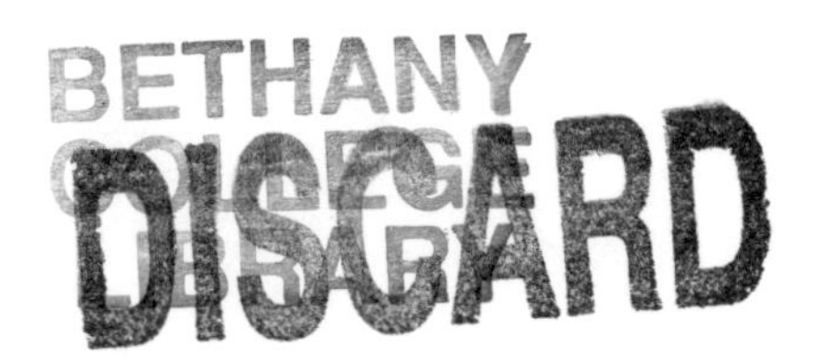
BETHANY
COLLEGE
LIBRARY
DISCARD